Environmental management at airports

Environmental management at airports

liabilities and social responsibilities

Edited by Nadine Tunstall Pedoe, David Raper and John Holden

 Thomas Telford

Proceedings of the international conference on *Environmental management at airports: liabilities and social responsibilities* held in July 1995 at Manchester Airport, organised by Scott Wilson CDM and Manchester Airport.

Published by Thomas Telford Publishing, Thomas Telford Services Ltd, 1 Heron Quay, London E14 4JD

First published 1996

Distributors for Thomas Telford books are
USA: American Society of Civil Engineers, Publications Sales Department, 345 East 47th Street, New York, NY 10017-2398
Japan: Maruzen Co. Ltd, Book Department, 310 Nihonbashi 2-chome, Chuo-ku, Tokyo 103
Australia: DA Books and Journals, 648 Whitehorse Road, Mitcham 3132, Victoria

A catalogue record for this book is available from the British Library

ISBN: 0 7277 2520 3

Typeset by Techset Composition, Salisbury, Wiltshire
Printed in Great Britain by The Cromwell Press, Melksham, Wiltshire

Preface

The last decade has shown an increase in public, government and industry concern around the environmental issues due to the operation of airports.

Aviation, like all industries, is facing the effects of increasing environmental pressure. Clearly, environmental pressures vary from country to country and from one airport to another, in part reflecting differences in social and political attitudes and also arising from historical events which may have heightened the awareness of a particular issue. Inevitably such pressures will continue to mount and will have a significant influence on the future development of airports worldwide.

The way in which individual airports deal with environmental issues will undoubtedly influence the development of aviation. As a consequence, it is imperative that there is a broad awareness among the world community of airports of environmental issues and solutions, and a recognition that collective expertise also facilitates the development of meaningful environmental legislation at national and international levels.

It is quite clear that if the aviation industry, and in particular, airports are to meet the predicted demand in air travel they must develop in a sustainable way. However, the industry is still coming to terms with sustainability and what it means for its business. Clearly, it does mean that the industry must be proactive and address more than legislative requirements. Indeed sustainability is interwoven with social responsibilities and it is often the way in which these responsibilities are addressed which determines the rate of growth an airport can achieve.

This book contains the papers presented at the International Conference on Environmental Management at Airports – Liabilities and Social Responsibilities, held in Manchester, United Kingdom, during 6–7 July 1995. The conference was organised by Manchester Airport plc and Scott Wilson CDM. The organising committee consisted of Peter Guthrie, John Holden and Colette Grundy of Scott Wilson CDM, and David Raper, Callum Thomas and Karen Wray of Manchester Airport plc. Thanks are also due to Fiona Brien, Angie Hall, Joanne Martin and Julie Wheatcroft for providing secretarial support, and to Gary Cooper, Louise Curme, Mark Webb and Jill Fisk for their work on the graphics.

The objective of the conference was to bring together senior managers from many sectors of the aviation industry to stimulate debate on environmental issues subject to legislative control and those which fall within the social responsibilities of Airport Authorities. The emphasis was on identifying current best practice and future constraints and developing an understanding of sustainability. Case studies were presented and discussed in detail.

The Table of Contents reflects the diversity of technical approaches used by the industry in developing solutions to environmental problems and in identifying issues likely to face the industry in the future.

The editors are grateful to the authors for their excellent contributions. We hope that this book will increase the understanding of the environmental pressures facing airports, identify current best practice and future constraints and add to the sustainable mobility debate.

The Editors
May 1996

Contents

Advancing environmental management through voluntary initiatives

Bob Ryder, Department of the Environment

The last few years have seen a strong development towards voluntary action by business to be proactive in managing and improving environmental performance. There has also been a growing recognition among governments that regulatory action – the setting and enforcing of minimum standards – is not enough on its own to deliver continual improvement across the whole range of performance. There is much to be gained, from government's perspective, through encouraging business to develop its own high standards of self-monitoring and management, with managers themselves owning both the responsibility and the systems for reducing environmental impacts and improving performance.

Increasingly, too, business has seen the commercial advantage of a proactive approach. The influence of public opinion, particularly through consumer markets, has never been greater. Companies at the leading edge, realising the commercial advantage, are in turn influencing the approach of their competitors and their suppliers. And, as well as being a factor in market share and income terms, performance in environmental management is also being seen increasingly as an opportunity to reduce costs and improve business efficiency.

The development of industry standards, setting the framework and quality specification for management systems, is also helping to raise public confidence in the ability of business to tackle environmental issues in a responsible and reliable way. BS 7750 has been a major advance in establishing the credentials of the voluntary approach. The European Community's Eco-Management and Audit Scheme (EMAS) is taking that process further, with specific recognition for manufacturing sites with established systems and with a commitment to real improvement which can be demonstrated through the publication of independently verified information. The future is likely to see extensions of the 'EMAS approach' into other sectors. And, for businesses which can demonstrate in this way that they are positively managing and delivering improvements in their performance, there are likely to be further shifts in the balance between regulation and self-monitoring.

One of the main themes in the programme for this conference is an understanding of the social responsibilities of airport authorities. Indeed, it is a key aim of the conference to draw out for senior managers in the industry how these responsibilities tie in with the goals of sustainability.

The notion of sustainable development has become firmly established since the 'Earth Summit' which took place in Rio de Janeiro in 1992. The UK was one of the first countries to follow up the summit with a national strategy for sustainable development – a programme of action to achieve the more sustainable global environment which national governments agreed at Rio. The basic aim is to reconcile the drive for continuing economic development (the drive for higher standards of living) with the urgent need to protect the environment from the effects of that economic growth.

This balancing act – to secure sustainability – is certain to increasingly affect, over the next decade or two, any industry involved in production or in the provision of services to meet consumer demand.

Governments on the whole will be taking the commitment to sustainable development very seriously, and will have a range of instruments to help them deliver that commitment. They have a number of broad options. One set of options relates to trying to change the nature of consumer demand.

☐ Governments can attempt this through programmes of education and public information, and some of the 'Agenda 21' initiative in the United Kingdom includes measures of this kind.

☐ Governments can also try to influence the behaviour of demand through economic instruments, either giving incentives to encourage environmentally helpful forms of activity, or using taxation to discourage forms of activity which have a harmful impact on the environment.

This debate about the potential for using economic instruments is of course particularly relevant to the demand for transport and leisure services, and will no doubt be familiar to those in the air travel industry. It is not proposed to pursue it any further here, but instead to consider another set of strategic choices which governments may make to help achieve sustainability. This other set of options raises some similar concepts about incentives and disincentives, 'carrots' and 'sticks'.

☐ To take the 'stick' approach first, a government, faced with an environmental issue, can choose a traditional route of additional regulation to achieve the effects it wants. It can try to reduce overall environmental impacts by legislating to raise minimum standards and by using its enforcement powers over industry to regulate compliance.

☐ The alternative approach is to encourage industry to take the initiative. Governments can promote voluntary schemes or agreements whereby industry is allowed fuller responsibility to manage its environmental effects to demonstrate its ability to deliver improved results.

This 'voluntary alternative' is not something which governments have traditionally been inclined to favour. But the tide is turning, and the change is likely to continue. There have been noticeable changes already, both in the philosophy of environmental groups and the approach of government and legislators. Until quite recently, western societies have often tended to be ambivalent, acquiescing in the benefits of economic and industrial development and being less fussy about social and environmental costs. That made it too easy, in response to particular environmental concerns, to blame industry for the problem, to legislate for stricter limits and to put more resources into enforcement.

Now there is a movement to a more honest dialogue on the environment – a better appreciation of the issues between industry, pressure groups and government. It is exactly the more realistic level of appreciation which is bound up in the whole concept of 'sustainable development'.

Industry is now increasingly seen, for example, as the provider of many of the solutions to environmental problems through the development of new processes and pollution control technologies.

More generally, even the most regulation-minded governments are appreciating that a simple reliance on regulation cannot guarantee results. Performance in environmental management, just as in any other branch of management, depends crucially on managers feeling responsible for the systems which they manage, and not simply following the demands imposed from outside the operation by a regulator.

The result is that policy makers in the UK and Europe are now starting to frame environmental legislation in such a way that it permits industry to show what it can do, rather than force it down a particular path. This has encouraged industry to adopt a more flexible response to environmental problems, one which is led and owned by company management rather than by government. It has also encouraged industry to move away from relatively short-term 'end-of-pipe solutions' to more fundamental 'process redesign' and the adoption of cleaner technologies.

One example of this approach is the European Community Ecolabelling scheme. This scheme provides a legislative framework which identifies products that are less damaging to the environment than equivalent brands across the whole of their life cycles. The key point about the ecolabelling scheme is that participation is voluntary: in order to achieve its environmental objectives, it uses the potential of market incentives – allowing industry to plug into the consumer demand for 'green products' – to encourage the design, production and marketing of less harmful goods and services.

The attitude of the market is in fact another key point in the changing picture. Whether expressed in terms of 'consumer demand' (which affects industry directly) or 'public opinion' (which affects governments more directly, as well as industry) the perceptions of consumers and voters about the importance of environmental issues have now become a critical consideration for any business. We have seen quite recently the impact which negative public perceptions can have internationally, bearing down on a single company over a single environmental issue. More positively, there are many examples of companies taking a proactive approach to the environment, and finding that there is a solid commercial advantage in being able to demonstrate that the environmental impacts of the business are being responsibly addressed and properly managed.

Manchester Airport is a good example of a business which takes environmental issues seriously and wishes, quite rightly, to be seen to do so. This corporate commitment by the Airport goes beyond the immediate issues of the site and its operations, and takes a wider view – in improving the environment of the Mersey Basin, for example, and other parts of the region. Manchester Airport also provides a good example of how a business can use its relationship with suppliers and customers in a constructive way to reinforce its management of environmental effects. Like an increasing number of major companies in the UK, it is following its environmental policies through into the supply chain and into the agreements which it makes with business partners.

A demonstrably responsible and positive attitude to environmental issues reflects the 'social responsibilities' theme of this conference, and is also a factor which pays off in terms of companies' market share and income.

The other theme of the conference – 'liabilities' – of course also has a financial pay-off. Good environmental management is increasingly being appreciated as an opportunity to reduce liabilities in the form of companies' costs – in the use of energy and other resources, and in the costs of waste and waste disposal (all of which is highly relevant to airport operations).

Good environmental management also reduces a company's liabilities in a more literal sense, in terms of the risk of statutory infringements and commercial penalties. If 'social responsibility' towards the environment can pay off in terms of market share, 'reducing liabilities' achieves its pay-off in the form of improved business efficiency.

The UK Government has been actively putting out these messages to corporate businesses through the promotion of voluntary schemes like the campaign for 'Making a Corporate Commitment', or 'MACC'. Manchester Airport is one of nearly 2000 major organisations in the UK which have signed up to MACC (indeed Manchester was one of the founder group of signatories) and to the underlying commitment of MACC towards the review and management of energy usage. The energy performance of these organisations is on average significantly better than for companies which are not part of the campaign and it is no coincidence that signatories to the MACC campaign are typically at the leading edge of business efficiency and productivity in the UK.

More recently, the UK Government has been putting a lot of effort into promoting the European Community Eco-Management and Audit Scheme, or 'EMAS'. This is a voluntary regime which has been developed alongside the growth of formal standards for environmental management systems. The development of such industry standards, which set a proper framework and quality specification for management systems, is going to prove extremely helpful in raising public confidence in the ability of business to tackle environmental issues in a responsible and reliable way.

The setting-up of BS 7750 has been a major advance in establishing the credentials of the voluntary approach. And the European registration scheme, EMAS, is now taking that process further, with specific recognition for manufacturing sites with established systems and with a commitment to real improvement – improvement that can be demonstrated through the publication of independently verified information.

Although EMAS does not at present formally extend to transport facilities and operations, that may change. The scheme is voluntary. It is site-based: companies wishing to participate will register their operations site by site with the national 'competent authority' (which in the UK is DoE). They may then use a special European symbol to demonstrate the achievement of their environmental management system and of their verifiable published information about environmental targets and performance. As well as a vehicle for delivering real environmental improvements, this is seen as having potential for very large marketing and business value for companies across Europe over the next few years.

The EMAS regime was originally framed by the European nations with manufacturing industry in mind, but there is scope for member states to extend it to other sectors. In the UK, EMAS has been extended to local government with the wholehearted support of the local authority sector. This will demonstrate how the eco-management and audit approach can be applied to much more diffuse operations such as the delivery of individual local government services – housing, planning, leisure and highways, for example. The future may well see extensions of the 'EMAS approach' into other sectors. The Department of the Environment, for example, is currently working on a pilot scheme to apply the approach to a major policy and spending area within central government.

There is no reason, therefore, why the airport sector should not follow the EMAS approach. There is already interest in certifying airport operation to BS 7750 and some airports, like Manchester, are already publishing environmental information and statements for an open, general readership. There would be a great deal of interest in a major European airport demonstrating that it is meeting all the essential requirements of the scheme. This would certainly raise the profile of environmental achievements both of the individual airport and the sector more widely.

This paper started by discussing the traditional regulatory approaches which governments have followed when trying to raise standards of environmental protection and performance, and how that philosophy is starting to change. Standards like BS 7750, and schemes like EMAS, which are rigorous, credible and can be independently tested and verified, are not only good for business and for reducing costs and liabilities. They are also an excellent way of demonstrating social responsibility in the management of environmental effects.

For those who actively participate, in what is, in effect, an open invitation from governments to help make the voluntary agenda work, there are some very interesting possibilities in the medium-term about the relationship with the various regulators involved. Company commitments to 'self-monitoring' and sound management will be an increasingly important factor in how regulators decide to deal with their responsibilities.

There are therefore some substantial opportunities now for businesses in all sectors to show that voluntary action to manage the environment and improve performance can be one of the driving forces in environmental policy over the next ten years.

Noise related to airport operations – community impacts

DR CALLUM THOMAS, Head of Environment, Manchester Airport plc, Manchester M90 1QX, UK

Airports need to ensure good relations with the residents of their surrounding communities if they are to sustain their development into the future. The disturbance caused by aircraft noise is the single most important environmental impact which local residents wish to see controlled.

Since noise disturbance is a form of nuisance and therefore a social issue associated with an individual's concept of 'quality of life', it is important that the airport's neighbours have an input into the design of the noise control programme. It is equally important for the Airport Company to set performance targets and that monitoring of, and adherence to those targets be publically reported and subject to third party auditing.

When an airport proposes a major development which will give rise to an increase in capacity there is often a widespread and very understandable fear among local residents about the implications of such growth for the local noise climate. In this context it is important that the airport considers offering a package of community guarantees about future environmental performance.

This paper explains how Manchester Airport has adopted this approach to the management of its noise impact.

INTRODUCTION

Manchester Airport is the fourteenth international airport in the world in terms of the passenger throughput. Last year Manchester handled more passengers than many other European Capital airports such as Athens, Copenhagen, Zürich and even Bruxelles. It is the third largest in the UK, after Heathrow and Gatwick and considerably larger than Leeds–Bradford and Liverpool, the other two main airports in North-

West England. The airport currently serves 170 destinations across five continents.

Over the past decade, Manchester Airport has consistently been one of the fastest growing airports in Europe. This year over 15 million passengers will pass through its terminals, arriving or departing on one of the 170,000 aircraft movements.

The demand for direct air travel into and out of the North-West of England continues and it is predicted that by the year 2005, over 30 million passengers will wish to use Manchester each year. However, the existing infrastructure will not permit the achievement of that target because of a lack of runway capacity. Manchester Airport is, in fact, already having to turn traffic away at the peak times of the day, simply because the runway is full.

In 1991, the Airport Company published its development strategy to the year 2005 which included proposals for the construction of a second runway. Those proposals were subsequently developed into a planning application which was made the subject of a Public Inquiry which lasted from June 1994 to March 1995.

The construction of a second runway would bring considerable benefits to the North-West of England, an area of high unemployment and social deprivation; benefits which are measured in tens of thousands of jobs and hundreds of millions of pounds associated with inward investment. While these benefits are felt across the entire region, the costs of our operation and future development are borne primarily by the residents of local communities.

Manchester Airport has a long tradition of addressing the concerns of its neighbours. In part this arises from the ownership of the Airport Company by the Local Authorities within the Greater Manchester area. Directors of the Company have a dual responsibility both to encourage the development of the Airport itself, and also to provide adequate protection to local people, many of whom are their electorate. The Airport Company also recognises that it will only be able to sustain its growth and development through effective management of the environmental impact of its operations. It is for these two reasons that Manchester Airport is striving to achieve a world class environmental management programme.

The control of noise is central to the company's environmental strategy. Noise disturbance is, however, primarily caused by aircraft, which are operated by the customer airlines. An effective noise control programme therefore requires the achievement of the correct balance between the needs of the customer airlines and the needs of the community.

This paper describes the noise control programme operated by Manchester Airport and the way in which the company has attempted to address the concerns of local residents about both current and future disturbance from airport operations.

CLARIFYING THE NOISE PROBLEM

In order to maintain an awareness of issues of concern to local residents and in particular to try to quantify the nature and extent of disturbance caused by the airport's operation, it is necessary to develop and maintain a suitable dialogue with the community. At Manchester we have done this through a number of different approaches.

The Manchester Airport Consultative Committee

The Manchester Airport Consultative Committee was set up in 1969 as the formal interface between the airport company and its neighbouring communities. The committee comprises 32 members representing local authorities affected by the airport's operation, amenity groups and user groups. It meets on a quarterly basis in public with the media present. It requests reports from the Airport Company on a variety of issues including the results of noise and track monitoring, developments in the noise control programme and an analysis of complaints from the community.

The work of the full committee is supported by a technical sub-committee comprising representatives of all key groups. This meets in private and is therefore able to discuss matters in greater depth. A variety of working groups have also been established to ensure that the most effective noise control programme can be achieved. This approach is the most effective way of ensuring that all groups can contribute to the noise control programme and enables the airport to take a greater role in chairing discussion, rather than acting as an intermediary between the various parties.

The Aircraft Monitoring Group acts as a conduit to introduce community concerns and wishes directly into the noise control programme. It guides the company on the priority of different issues to the community and can monitor and report progress on the noise control programme to the full committee. The Pilot Technical Working Group comprises pilots from the major airlines operating at Manchester, representing a wide range of aircraft types. The group enables the practical experience of pilots and aircraft operations to be used to guide noise control policies. An important secondary role is to enable pilots to talk to each other, to share ideas and improve operational practices.

The Environmental Health Officers Consultative Group comprises EHOs from local authorities affected by our operations. This introduces a further area of expertise and an independent authoritative dimension into the process. It allows direct consultation and is an audit vehicle where detailed information can be tabled and discussed. A similar body is currently being established to enable consultation with Local Health Authorities.

Social surveys

One way of discovering how the airport affects the lives of local residents, and what action they would like to see taken to alleviate that impact, is to seek their views. In 1988, a social survey was commissioned in nine communities in the vicinity of the airport and its flightpaths, and a tenth control community for a site unaffected by aviation (in Bolton). The results of the survey were published through the Airport Consultative Committee and formed the basis for the development of an improved environmental control programme. Although Manchester Airport has operated a noise control programme for over 20 years, the survey revealed that disturbance caused by aircraft noise was still clearly the single most important concern of local people.

The survey revealed that while most people were disturbed by aircraft during the evenings, when relaxing or putting their children to bed, disturbance to sleep was seen as the most unacceptable intrusion associated with the airport's operation.

While there was considerable support for Manchester Airport, even in the most affected areas, over half of those interviewed believed that the airport was not doing enough to reduce noise disturbance. Generally, while people were willing to tolerate some degree of disturbance as inevitable, they did object to that which could be avoided. This was seen to arise from two sources:

☐ night flights by Chapter 2 aircraft
☐ aircraft deviating from flightpaths.

There was a widespread belief that it is difficult, if not impossible, for the airport to reconcile its commercial interests with its responsibilities to the local community. Many people believed it was not worth complaining to the airport and were unhappy about the way in which complaints were handled. Some individuals believed that when they were disturbed at home they had the right to expect that the airport would take their complaint seriously and act upon it. They were not convinced, however, that this was happening.

The establishment of an Environment Department within the Company in 1992 was an attempt to address these concerns. A Community Relations Unit was established within the department with specific responsibility for identifying issues of concern to local residents and assisting in the development of measures to mitigate the impact of airport operations upon the community. The unit handles complaints from residents, investigates them and acts as a voice on behalf of the community within the Company.

Complaints from local residents

Manchester Airport has maintained a community complaints investigation and reply service for many years. A noise complaints telephone line is maintained and the number published. The telephone line is staffed during normal working hours with an answerphone service at night and at the weekend, although weekend cover is provided during the busy summer period. At all hours, day or night, people can also speak to an Airport Duty Manager.

All complaints received are logged, investigated and the results analysed and reported to senior management and the Airport Consultative Committee. Complaints relating to specific movements or operations are taken up directly with the airline concerned. Specific reports on operations, or operators which give rise to significant numbers of complaints are prepared for discussion by the Senior Management Team.

Complaints indicate the types of aircraft which cause most disturbance. These are principally slow climbing aircraft such as Jumbos (B747s) and aircraft which create most noise (Fig. 1).

The total number of complaints recorded by the Community Relations Unit in 1993 was 4747. Two-thirds of these related to operations by

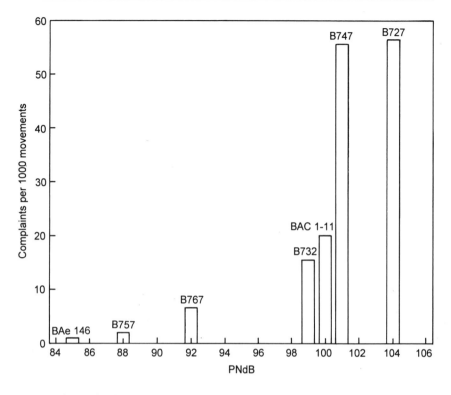

Fig. 1. Relationship between the noise level of the departing aircraft and the number of complaints received per 1000 movements

Chapter 2 (noisy) aircraft despite the fact that they comprised only 17% of movements in that year. The proportion of complaints relating to poor track keeping has increased progressively in recent years from one-third in 1990 to two-thirds today. This in part relates to an increased awareness of where people expect to see aircraft flying, but may also reflect the declining numbers of movements made by the noisier types of aircraft.

Individuals vary considerably in what they believe to be disturbance and their propensity to complain. Of 4747 complaints registered in 1993, 2153 came from only 10 addresses. The most frequent complainant registered 987 complaints in a year and the second most frequent 656. The actual number of complaints may not, therefore, be a good indicator of the level of disturbance in a particular community. Rather, it may reflect the views of a small number of individuals. The total number of people who complained in 1993 was 987.

Table 1. Total number of noise and track keeping complaints registered by the Community Relations Unit in recent years

Year	Number of complaints
1989	2042
1990	2021
1991	3280
1992	3601
1993	4747

The number of complaints received by the airport in recent years has increased as illustrated in Table 1.

This increase has probably occurred for a variety of reasons:

☐ the growth of the airport
☐ changing public attitudes and media interest
☐ the level of organised opposition to airport growth
☐ publicity associated with the airport's environmental programme
☐ publication of the airport's development proposals
☐ responses from the public to specific issues of concern
☐ the sustained efforts of a small number of persistent complainers.

The geographic distribution of complainants relative to departure and arrival routes for the airport is illustrated in Fig. 2. The number of complaints from different communities changes from year to year and can highlight issues of local concern for action. While the total number of complaints may not be a good indicator of the level of disturbance in a particular locality, complaints are very useful for identifying areas for investigation and possible action by the Environmental Control Unit.

CURRENT NOISE CONTROL PROGRAMME

The Company's current noise control programme, which is designed specifically to reduce the level of disturbance to the local communities, contains the following key elements.

☐ The involvement of community groups, Local Authority Environmental Health Officers (EHOs), airlines and the National Air

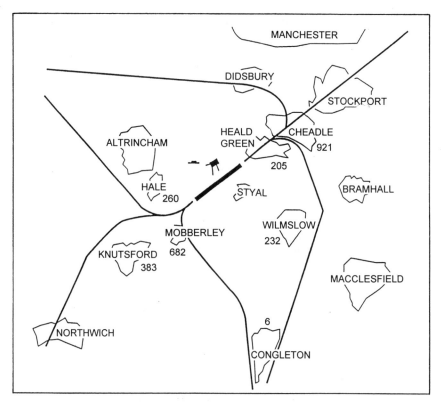

Fig. 2. Geographical distribution of complaints relative to airport departure routes

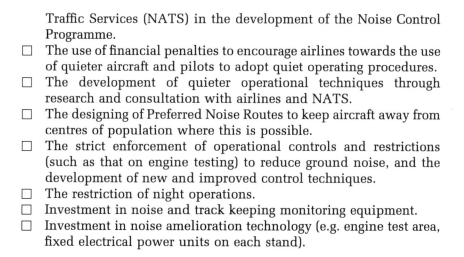

Traffic Services (NATS) in the development of the Noise Control Programme.

☐ The use of financial penalties to encourage airlines towards the use of quieter aircraft and pilots to adopt quiet operating procedures.

☐ The development of quieter operational techniques through research and consultation with airlines and NATS.

☐ The designing of Preferred Noise Routes to keep aircraft away from centres of population where this is possible.

☐ The strict enforcement of operational controls and restrictions (such as that on engine testing) to reduce ground noise, and the development of new and improved control techniques.

☐ The restriction of night operations.

☐ Investment in noise and track keeping monitoring equipment.

☐ Investment in noise amelioration technology (e.g. engine test area, fixed electrical power units on each stand).

☐ The provision of sound insulation in the surrounding communities.

☐ Agreement and securing of policies in Local Plans in surrounding areas.

☐ Comments on and objections to noise-sensitive development proposals in areas subject to higher levels of aircraft noise.

Monitoring

The noise control programme is underpinned by the Company's aircraft noise and track monitoring system, known as MANTIS (Manchester Airport Noise and Track Information System).

The system comprises a central computer connected by telephone lines to eight fixed monitoring points at strategic locations in the surrounding communities and on departure and arrival routes. An additional four mobile monitoring points can be placed anywhere within the vicinity of the airport (usually at the request of local residents or EHOs) and transmit noise monitoring data to the computer by cellphone.

Radar data which is provided by NATS is input directly into the system. This information reports both the height of the aircraft and its geographic location.

The Airport Company's Flight Movement Computer System feeds details of the aircraft into MANTIS. These include its callsign, the airline, the type of aircraft and the departure route it will follow.

The Air Traffic Control Services enter details of any aircraft which has been given permission to deviate from the planned departure route and follow a Non-Standard Departure (NSD).

The system will soon also record three radio channels on which Air Traffic Controllers talk to pilots on departure. This will allow the airport to investigate rapidly an aircraft which deviates from its departure routes.

The MANTIS system (developed in part at Manchester) is used to:

☐ penalise noisy, and eventually penalise off-track aircraft
☐ develop better noise abatement and track keeping procedures for aircraft

☐ measure the effectiveness of the noise control programme
☐ describe the current noise climate
☐ allow the correlation of complaints and noise events
☐ allow the production of regular monitoring reports.

The Company is currently investigating methods of enabling direct access to the system for operations staff, the air traffic control services and local authority EHOs. An open door policy already exists to enable these groups to access information during office hours.

The community complaints handling service is being integrated into the working of the MANTIS system, as the majority of complaints relate to noise or off-track aircraft. This enables a complaint to be rapidly matched to a particular aircraft movement. MANTIS also has the capability to identify and record automatically all aircraft which fly off-track or exceed particular noise levels. It is this information, rather than complaints, which is used to enable environmental penalties to be applied to airlines and the effectiveness of the noise control programme to be assessed.

The move towards quieter operation

Changes in aircraft type and improvements to operational procedures have resulted in a significant drop in noise levels made by departing aircraft operating out of Manchester over the past 14 years. This is illustrated in Fig. 3. The average peak noise level made by departing aircraft has declined from approximately 106 PNdB in 1971 to 95 PNdB in 1993.

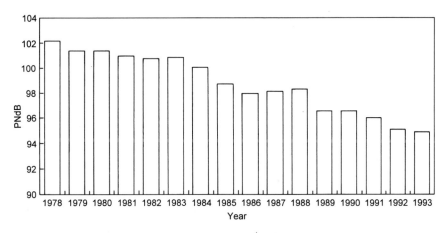

Fig. 3. Average peak noise level of departing aircraft 1978–1993

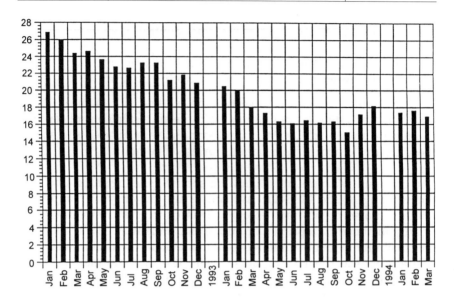

Fig. 4. Percentage movements made by Chapter 2 aircraft: Jan. 1992–Mar. 1994

In the past two years there has been a rapid decline in the percentage of Chapter 2 aircraft operating out of Manchester, as shown in Fig. 4.

These improvements have been encouraged by various noise penalty and landing fee rebate schemes operated by the Airport Company. A new penalty scheme came into operation in 1994 in which the daytime threshold was lowered by 5 PNdB. This involves a minimum penalty of £500 for aircraft which exceed 105 PNdB by day, or 100 PNdB at night (23:01–06.59 hours) as measured at a point 3.5 nautical miles from the start of roll, with an additional escalating penalty in proportion to the level by which the threshold was exceeded. Monies raised from noise penalties are not retained by Manchester Airport but are used to provide grants for projects which enhance the 'quality of life' within communities affected by the operation of the airport.

The Community Relations Unit handles requests for support on sponsorship from local organisations. The unit currently has an annual budget of £50,000 which is increased if noise penalties exceed that figure. In 1993/94, support was provided for a range of projects including:

☐ Woodhouse Park Youth and Community Centre Playscheme
☐ St Ann's Hospice

- ☐ Manchester Junior Chamber of Commerce
- ☐ Alder Hey Hospital Fun Book
- ☐ supply of computer software to the *Big Issue*
- ☐ Schools Speech and Drama Festival
- ☐ Wythenshawe Pensioners Association Christmas party
- ☐ Quarry Bank Mill – audio tape tours for visitors
- ☐ Vale Royal Enterprise Agency – export project.

The work of the Pilot Technical Working Group and research by Manchester Airport has resulted in the development and adoption of improved takeoff and landing procedures through education and by encouraging the sharing of expertise.

The Company is currently undertaking a systematic review, with airline assistance, of all noise-related operational techniques employed across European airports with a view to achieving further reductions in noise emissions from air traffic at Manchester and to ensure we are adopting best practices.

A system of noise related operational charges has recently been introduced in order to further hasten the date when all movements are made by Chapter 3 aircraft.

Aircraft tracking

The airport currently operates a system of six preferred noise routes (PNRs) designed to take departing aircraft as far as possible away from built-up areas, as illustrated in Fig. 5.

The detailed design of the PNRs is determined by a number of factors as well as the location of population areas. These include the airspace available to aircraft using Manchester Airport, the complexity and distribution of ground navigational aids in the area, areas of high ground, international safety regulations and the performance characteristics of different types of aircraft.

Departing aircraft are required to stay on the PNRs until they reach an altitude of 3000–5000 feet. The height at which aircraft are released from the PNRs has been changed in the past in consultation with the local community in order to achieve a reduction in the overall level of disturbance caused along a particular route.

Whenever changes to PNRs are considered, they are always the subject of extensive public consultation. In the past, this has led to proposed

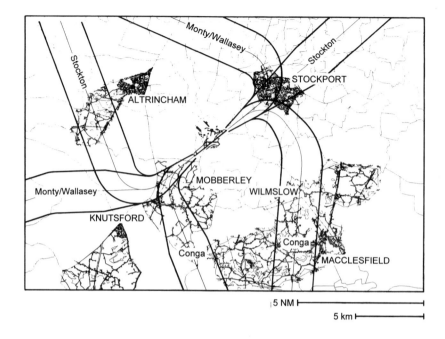

5 NM ├──────────────────┤

5 km ├──────────────────┤

Fig. 5. Current arrival and departure routes (copyright licence number AL18020A)

changes not being implemented because of lack of public support, despite there being potential noise benefits in terms of the number of people affected.

In 1987, a number of changes were introduced to the design of the PNRs for departing aircraft. At the same time as these changes took place, a heightened awareness of aircraft tracking issues developed in the local communities and there then followed a period of a year when the airport received an increasing number of complaints about 'off-track aircraft'. As a result, in 1989 Manchester became the first airport in Britain, and one of only five in Europe, to install an aircraft track monitoring system.

This offered, for the first time, the opportunity and ability to system-atically research and monitor the degree of accuracy of track keeping along each of the PNRs. This work was undertaken with the close involvement of members of the Consultative Committee, airlines and the National Air Traffic Control Services. It included:

☐ consideration of revisions to the design of PNRs to improve the accuracy with which they can be flown and to reduce the number of people overflown

☐ action to enable airlines and pilots to develop improved operating techniques
☐ increased constraints upon the freedom of operation of air traffic controllers (ATCOs)
☐ an awareness programme for pilots and ATCOs.

Tracking penalties

Members of the Airport Consultative Committee considered that airlines which failed to adhere strictly to the PNRs should be penalised. The Senior Management of the Airport agreed and, following consultation with airlines at Manchester, the Airport Board approved the introduction of a track penalty scheme.

Following a period of negotiation with the Airline Operators Committee at Manchester a financial penalty scheme was introduced in 1992. A 50% surcharge on landing fees was proposed as a track penalty to be introduced as part of the overall schedule of fees and charges to airlines operating at Manchester.

Soon after issuing the first set of surcharge invoices, the International Air Transport Association (IATA) wrote suggesting that further discussion should take place before such a scheme could be implemented.

IATA challenged the legal basis for charging a supplementary landing fee and the calculation of PNR track tolerances. In addition, concern had been expressed that legitimate deviations from track should not be penalised. The Company, following further legal advice, agreed that it would withdraw the invoices pending further consideration of the issue. It is believed that IATA's concern arose from the fact that Manchester had become the first airport in the world to introduce track penalties and that it would set an important precedent.

The intervention by IATA resulted in Manchester Airport deferring the introduction of tracking penalties until legislative support could be achieved and also an agreed definition of how far an aircraft could deviate from the centreline of the PNR before it was considered 'off track'.

Accordingly we approached the UK Government's Department of Transport in order to achieve the necessary legislation as part of our response to its August 1991 Consultation Paper on the control of aircraft noise.

In March 1993, the DoT published the document 'Review of Aircraft Noise Legislation – Announcement of Conclusions' which contained the following commitment:

> The Government proposes to introduce a general enabling power giving aerodromes explicit powers to prepare noise amelioration schemes and to penalise operators who do not comply with them.

We are still awaiting the necessary window within the parliamentary legislative programme. The Airport, together with the Consultative Committee has since pressed the Secretary of State to progress this matter. Additional pressure has also been provided by Local Authorities, local MPs and representatives of local amenity groups.

In the absence of the ability to penalise off-track aircraft the Company will continue to write to all airlines whose aircraft transgress the PNR boundaries. Airlines are required to offer an explanation for the event. Finally, the airport has published and will continue to publish the names of those airlines that receive deviation letters through the Consultative Committee and make the information available to the press.

Our initiatives on this subject have led us to encourage the setting up of a European Airports Working Party and to actually establish one for UK airports. The aim is to agree expected standards of aircraft performance and a definition of the width of a departure track.

Accuracy of tracking by different types of aircraft

Different types of aircraft vary in their ability to follow the centreline of the PNRs. As a general rule, modern aircraft tend to be more accurate than older (Chapter 2) types.

There is considerable potential for increasing the accuracy of track keeping in the future through the reprogramming of the flight management computers installed in modern aircraft.

Research has revealed that differences in tracking accuracy by B757 aircraft flown by different operators can be traced to the make of software each company uses. Trials undertaken at Manchester with an operator of B737-300s have shown the benefits of such reprogramming in the same aircraft, as can be seen in Fig. 6. In the figure, each line represents the track followed by a departing aircraft. The two heavy lines represent the boundary of the ±1.5 km track. Discussions are

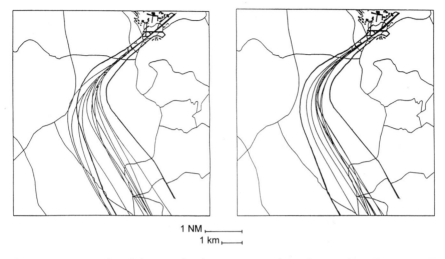

1 NM ⊢————⊣
1 km ⊢————⊣

Fig. 6. Accuracy of track keeping by departing aircraft can be significantly improved by reprogramming on-board navigational guidance computers

now underway with the airlines flying out of Manchester and NATS to assess the potential for all operators of such aircraft to consider this approach.

The increased use of the on-board navigational equipment or flight management system (FMS) available in modern aircraft will enable more accurate track keeping in the future. Unofficial trials have been carried out with airlines at Manchester, but the concept of FMS-based departure routes had not yet been accepted as standard practice in the UK by the CAA. At the present time, therefore, standard procedures cannot be published for FMS-based SIDs. Following recent discussions with the CAA, however, the first official trials of FMS standard instrument departures in the UK will soon begin at Manchester. These will involve the Airport Company, airlines, the CAA and NATS.

Improvements in track keeping

The success of the approach adopted at Manchester Airport may be assessed from the fact that the number of 'off-track' aircraft (that is the proportion of aircraft which deviate outside the ± 1.5 km track before reaching the transition altitude) has fallen from around 13–16% in 1992 to approximately 8–10% today against the same tolerances.

Aircraft are sometimes seen operating away from the departure tracks if they have been given 'a non-standard departure' or NSD by the Air

Traffic Controller. This may be for safety reasons, to avoid conflict with other traffic, to avoid bad weather, or at busy times to enable the free flow of air traffic. The frequency with which the National Air Traffic Services issue non-standard departure instructions has declined markedly over the past two years. This is partly due to an awareness campaign amongst air traffic controllers, and partly due to the fact that all NSDs now have to be logged by the Air Traffic Control Officer (ATCO) and entered into the MANTIS system. In addition, NSDs may not now be issued at night, neither can they be issued at any time to jet freighters, unless there is a safety reason for doing so.

Night flying

In recognition of the particular disturbance caused by night flying, the airport commissioned a study into the effects of aircraft noise upon sleep in 1988. This was carried out by Dr Ken Hume of Manchester Metropolitan University alongside the airport's social survey. This work contributed to the development of the major research project undertaken by Dr Hume and others for the DoT between 1991 and 1993.

Manchester Airport has operated a night flying policy (which relates to the period 23:30–06:00), broadly similar to that introduced by the DoT for the designated London Airports, since 1978. It comprised a fixed summer night quota with an apportionment of movements to different types of aircraft according to how much noise they produce.

Research and complaints both at Manchester and elsewhere had shown that it is not the frequency of traffic but rather the noise levels of individual movements which causes disturbance at night. With this in mind, the Company's night flying policy aimed to achieve a rapid elimination of movements made by the noisier types of aircraft whilst allowing traffic growth to occur using quieter models.

Progress in the elimination of movements by noisy jet aircraft at night may be assessed from Table 2. The number of planned night movements by such aircraft declined from 4855 in 1986 to only 955 in 1993.

The Company has just adopted a new night flying policy for the years 1994–2005. This applies a dual control: a movement limit and a noise budget.

Each night movement counts against the budget. Noisier aircraft 'cost' more points. Quieter aircraft 'cost' less.

Table 2. Recent changes in the number of movements scheduled to be made by noisy jet aircraft at night between 1986 and 1993

Year	No. of slots
1986	4855
1987	4581
1988	3486
1989	2925
1990	1920
1991	1715
1992	1515
1993	955

The budget has been set to the figure that would have been achieved if all movements in the 1992/93 night period had been operated within this regime. That is to say, all night movements in the year 1 November 1992 to 31 October 1993 were analysed, allocated a point score according to the type of aircraft operated, and an annual total obtained.

The objective of the policy and associated operational controls is to ensure that night flights (between the hours of 23.30 and 06.00) do not create more disturbance, or exceed the noise levels recorded in 1992/93 or the area of modelled noise contours derived from the 1992/93 aircraft movement rates and noise exposure.

Preferential runway use

In recognition of the fact that the largest number of people live to the north-east of the airport, the Company has adopted a policy of making preferential use of Runway 24 for departures. Thus aircraft take off over the less densely populated areas.

Ground noise

About 5% of complaints received by the airport can be traced to ground noise. In 1993, this totalled approximately 240 of which 180 came from a single individual, who lives one mile to the north of the runway.

Ground noise can be kept to a minimum in a number of ways. The use of fixed electrical power units to provide electricity to aircraft on stand, as opposed to use of the aircraft's auxiliary power unit (APUs), provides some relief. The minimal use of reverse thrust by landing aircraft, especially at night, is also important and is being addressed with the airlines.

A recent audit has revealed that on average, aircraft on stand operate their auxiliary power units for approximately 45–60 minutes on each turn around.

However, aircraft can also receive electricity from grid supplied fixed electrical power units (FEPs). These have been supplied to most stands and the target is to increase the use of FEPs by $33\frac{1}{3}$ percent by 1998 over the 1992 level.

New taxiing and start up procedures have been investigated as a result of a series of complaints from a resident in Hale Barns. This has involved turning aircraft 'nose out' on stand and the preferential use of aircraft stands which are further away from the village.

When the Company recently approved construction of a new aircraft maintenance hangar, it insisted that a soundproof engine test area be provided as part of the project. Prior to that all tests had been carried out on the open airfield. Accordingly, an engine test facility was constructed in 1992. Controls and monitoring systems are now in force to ensure that disturbance from engine testing is minimised. A key element of the control on engine testing is a ban on engine testing at night between 23:00 hours and 06:00 hours except in emergency situations.

The above measures designed to reduce ground noise on the airport itself will bring greatest benefits to the residents of the nearby communities of Hale Barns, Wythenshawe, Moss Nook and Ringway.

Sound insulation grants

In order to reduce noise levels in houses close to the airport and departure and arrival routes, the Company has maintained a sound proofing scheme since 1972.

The Manchester scheme makes provision for the installation of secondary sound insulation in domestic properties within an area derived from the 62 LAeq (24 hour) noise contour. The boundary has been extended in some areas to take account of natural features such as rivers or open land so as to avoid the situation of close neighbours falling on either side of the boundary.

It has also been extended to take account of landing traffic over Stockport. In addition, in recognition of the anticipated success of our efforts to improve aircraft tracking, the scheme has been extended along each of the preferred noise routes as indicated in Fig. 7.

All domestic properties within the boundary are eligible to receive a grant covering 80% of the cost of installing secondary glazing in five rooms up to a current maximum figure of £2400. It is estimated that there are approximately 15,000 domestic properties within the scheme. Provision is made for 100% grants in case of hardship and discretionary awards may also be made in special circumstances.

Schools, hospitals and residential homes have received discretionary grants for sound insulation in the past. The Company currently earmarks a budget of £1,000,000 per annum for the Sound Insulation Grants Scheme.

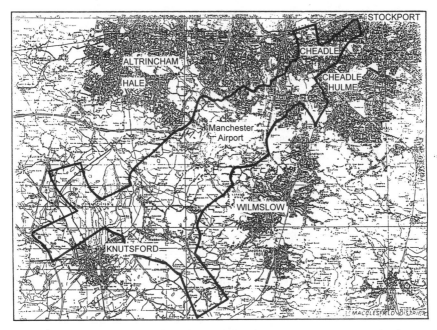

Fig. 7. Manchester Airport Sound Insulation Grants Scheme (copyright licence number ALI8020A)

FUTURE DEVELOPMENTS

In 1991, the airport published its draft Development Strategy to 2005, which included details of the research into three potential sites for the construction of a second runway. Extensive public consultation followed which included public meetings and exhibitions in the areas which would be affected, and the production and distribution of 10,000 brochures which incorporated a questionnaire designed to elucidate the attitudes of local residents.

The questionnaire returns again confirmed that aircraft noise impact was the single most important issue of concern to local people. Furthermore, the results revealed that there was genuine concern (or fear) about the effects of growth and what future quality of life would be like in communities affected by our operations if a second runway were constructed.

It was for exactly this reason that the airport had raised the concept of offering community and environmental guarantees to underpin the second runway when it first published its draft Development Strategy proposals. These guarantees were to take the form of mitigation measures, environmental targets and commitments. They would be designed to address the key issues identified by the Environmental Assessment and public consultation undertaken as part of the second runway planning application process.

Following submission of a formal planning application, Cheshire County Council, one of the two planning authorities within whose boundaries the airport lies, adopted a position of recognising the considerable benefits which would accrue from the second runway development but expressed objection to the development over the potential environmental impact.

Accordingly, following a period of negotiation, an agreement was drawn up which provided an acceptable level of protection and additional mitigation measures and controls, sufficient to permit Cheshire County Council to support the runway development proposals. These conditions were drawn up into a Legal Planning Agreement.

Appendix A provides a summary of the key elements contained within the agreement which relate to the control of aircraft noise disturbance.

The agreement is designed to reassure the neighbouring communities that the airport will continue to implement a comprehensive management programme for controlling noise. It contains legally binding commitments with a target to ensure that the noise climate will not deteriorate beyond that measured in 1992–1993 either during the day or at night, regardless of the growth of the airport, until at least the year 2011.

The agreement will establish what will probably be the most comprehensive and sophisticated noise control programme of any major UK airport. The most important components are:

☐ the setting of clear and measurable targets
☐ the maintenance of a suitable noise and track monitoring system
☐ regular publication of monitoring results
☐ third party auditing of systems and performance
☐ extensive and widespread consultation
☐ regular reviews
☐ action to reduce noise at source
☐ action to ensure the minimum number of people are exposed to any disturbance caused.

As such, the approach embodies the key elements of an effective environmental management system.

The Noise Control Programme would be underpinned by a continuing commitment to the Sound Insulation Grants Scheme.

To be fully effective, the agreement will require the co-operation of airlines and the National Air Traffic Services. This will necessitate continuing and close liaison between all three bodies.

There is a belief (which was expressed by some local residents in the 1988 social survey) that in the relationship between an airport and its customers the airlines, commercial expediency will always override environmental commitment. The obligations detailed within the S106 agreement are legally enforceable ones and require a corporate approach to the achievement of environmental targets. They cannot be achieved by staff from the Environment Department alone. For the first time, managers throughout the Company from Airfield Operations through Financial Regulation to Market Development all have specific responsibilities and have been assigned management targets to ensure that the conditions of the agreement are met.

Subject to the granting of planning permission for a second runway, Manchester Airport will be obliged to adopt a system of noise control which is publicly accountable. It will publicly set targets and model anticipated performance and will publicly report actual performance against target. Finally, the development of the noise control programme will be subject to extensive public consultation.

Systems and performance will be subject to frequent review and third party auditing with a brief requiring the auditor to recommend improvements. This in itself will encourage the achievement of best practice.

In general, the commitments given in the S106 agreement aim to secure an agreed level of noise protection for local residents until at least the year 2011. At that time, the parties to the agreement will endeavour to reach agreement on any amended proposals in respect of most of the obligations. In the event that this cannot be achieved, the existing obligations will remain in force.

CONCLUSIONS

The control of disturbance caused by aircraft noise, unlike many other aspects of environmental management, is a social issue which must be handled as such. The degree to which individuals are disturbed by noise is a matter of perception, as is their willingness to tolerate the disturbance.

Most people living close to an airport are willing to accept a reasonable degree of disturbance, as the Manchester Airport social survey revealed. However they do find certain types of aircraft operation totally unacceptable, either because the aircraft is extremely noisy, is not following published arrival or departure routes, or because it is perceived to be unsafe.

Since noise control is designed to minimise the disturbance caused to the airport's neighbours, it is important that they have adequate participation in any management programme. An effective noise control programme should:

☐ be designed and implemented with the active involvement of the community
☐ be meaningful to local residents and address the issues of concern to them
☐ include clear and measurable targets which are published
☐ incorporate a monitoring and reporting system which is credible to local people.

Aircraft noise, however, arises from operations made by airlines (the airport's customers) and on behalf of the general public. It is generally true that the airports which are most attractive to airlines are those which are close to the largest urban conurbations, those same airports which are, by definition, the most noise-sensitive.

Given the fact that airports are generators of regional development and that air travel will play an ever-increasing role in global transportation, it is in the interest of all to ensure that noise disturbance is kept to a practical minimum. It has fallen to airports to be at the centre of such issues. An airport has to achieve a very careful balance between the needs of the region it serves, the needs of its customers and the needs of the local community.

Achieving the correct balance is not easy, nor will it become any easier in the future. As society becomes more affluent, the demand for airport capacity will increase as more people want to fly, either for business or for pleasure. Those same people will, however, be less tolerant of pollution or disturbance from aircraft noise. This a paradox which airports need to resolve if they are to continue to develop and grow.

APPENDIX A: OBLIGATIONS AND GUARANTEES GIVEN IN THE SECTION 106 AGREEMENT WITH CHESHIRE COUNTY COUNCIL

Sound Insulation Grant Scheme (SIGS)

The Airport Company will continue to provide noise insulation for domestic properties.

- ☐ The SIGS boundary will be derived from 62 LAeq (24 hour) contour.
- ☐ An annual review of the conditions of the SIGS will be made.
- ☐ Boundary extended to take account of Runway 2.

Noise control

The noise impact of the airport's operation shall be no worse than that measured in 1992, as measured by:

- ☐ maximum noise levels of departing traffic (average of the noisiest 10% of movements)
- ☐ the modelled area of the 60 LAeq (07:00–23:00) noise contour.

Maintain a suitable noise and track monitoring system comprising:

- [ ] four new fixed monitoring points
- [ ] guaranteed access to the system for the Environmental Health Officers Consultative Group, the Airport Consultative Committee and the Health Authority Consultation Group
- [ ] annual external independent auditing of the noise impacts
- [ ] auditing of the accuracy and consistent operation of the system
- [ ] the production of noise contours for the previous year
- [ ] the production of forecast controls for the forthcoming two years
- [ ] monthly Noise and Track Monitoring Reports
- [ ] monthly community complaints monitoring and analysis reports.

A 'quietest operations policy' entailing the following:

- [ ] phase-out of Chapter 2 operations ahead of legislation
- [ ] the attainment of the following targets:

    - 100% scheduled night movements by Chapter 3 aircraft by 31/12/96
    - 92% scheduled operations by Chapter 3 by 1998
    - 96% scheduled operations by Chapter 3 aircraft by 2000

- [ ] monitoring of performance against targets
- [ ] operational penalties for aircraft breaching agreed noise limits
- [ ] annual review of the noise levels at which penalties applied
- [ ] developing a noise related element to landing charges
- [ ] the preferential use of Runway '24'
- [ ] annual review of opportunities for preferential use of single runway operations on the existing runway.

Ground Noise Control Policy

- [ ] No more than 20 engine tests between 23:00–06:00 hours within the engine test bay.
- [ ] No engine tests between 23:00–06:00 outside the test bay.
- [ ] Engine testing will commence in the engine test bay.
- [ ] A target to increase use of fixed electrical power units by $33\frac{1}{3}$% by 1998 over 1992 levels.

Preferred noise route and aircraft track keeping policy

- [ ] PNRs designed to avoid aircraft overflying built up areas.
- [ ] PNRs flyable by all types of aircraft.
- [ ] Systems to enable improved accuracy of track keeping and monitoring.

☐ Changes to noise abatement height of each PNR to be regularly reviewed.

☐ No change of PNRs without prior public consultation.

☐ Target PNRs to be ±5° either side of the track centre line.

☐ All aircraft operating outside the ±1.5 km track to be monitored, reported and investigated.

☐ In 1998 and 2002 an assessment of the ±1.5 km track width to be carried out to maintain progress towards a ±5° tolerance.

☐ Track keeping accuracy to meet the following target:

  – maximum of 5% of standard instrument departures off track by 1998 (1.5 km limit)
  – maximum of 5% non-standard departures by 1998.

☐ To introduce financial penalties for off-track movements as soon as statutorily empowered to do so.

☐ Special studies to be undertaken into localised refinement to departure procedure and PNRs.

☐ By December 1994 – to review departure procedures for special events at Tatton Park.

☐ By 31 December 1994 – to review the PNRs passing over Tatton Park.

Night flying

The noise impact of night flights up to 2005 will be no worse than that measured in 1992/93, as measured by:

☐ maximum noise levels of departing aircraft (average of noisiest 100 movements)

☐ the modelled area of 60 LAeq night noise contour.

Achieve 100% of scheduled night movements by Chapter 3 aircraft by 31 December 1996.

Adopt and implement a Night Flying Policy to include a points budget and movement limit to cover the period 23:30–06:00 hours.

☐ Total movements scheduled during the night shall not exceed 7% of total movements.

☐ All non-exempt aircraft will count towards the points budget.

☐ Movements limit will exclude aircraft which are demonstrably operated below 87 PNdB as measured at agreed monitoring points.

☐ No slippage of points between years or seasons.

Between 22:00 hours and 06:00 hours preference shall be given to using the existing runway.

Additional night time restrictions.

☐ Those types of aircraft which may be scheduled to land or take off between 23:00 and 06:00 hours.

☐ Lower threshold for noise penalties between 23:30 and 07:00.

☐ Non-standard departures not normally allowed between 23:00 and 07:00.

☐ Minimum use of reverse thrust between 23:00 and 07:00.

Noise abatement at Zürich airport

ANDREAS MEYER, Aircraft Noise Abatement Department, Zürich Airport Authority, PO Box 1513, CH-8058 Zürich, Switzerland

The growth of air traffic after the Second World War was based on a deep belief in technology and the enthusiasm to conquer the third dimension. Aircraft noise around airports was a major potential impediment to growth and noise abatement was identified as being an important technical issue. Zürich Airport was no exception in this respect, however, in comparison with other airports, noise abatement measures were implemented at a much earlier date. The beginning of the jet age at Zürich Airport was 1964 and thereafter there was a high growth in the percentage of such aircraft. The first noise abatement employee was hired at this time, and in 1966 a simple noise monitoring system was installed. This was due to the nature of Switzerland's political system; as the airport belongs to the Canton of Zürich, all important decisions, including financial decisions, have to follow political procedures, based on referendums.

Against this background, noise abatement had, and still has, the function of a form of political legitimisation. This paper discusses the political influences, describes the relations between the airport and the surrounding communities, and gives an overview of the most important technical mitigation measures and their results.

INTRODUCTION

Berthold Brecht wrote a play entitled *The exception and the rule*. In a number of ways Switzerland is an exception. The way in which conflicts between airports and their neighbours are dealt with is unique in many respects. However, Switzerland also follows the rule that the conflicts hinge, as they do in other countries, on aircraft noise.

This paper first outlines some of the unique aspects of the situation in Switzerland and the consequences for airports, particularly those at Zürich.

☐ Zürich airport was opened in 1953, and in 1964 the first noise abatement officer began work. Zürich was, therefore, one of the pioneers in this area.

☐ Since 1946, a series of 13 referendums concerning the airport or questions of air traffic has been held, of which two were negative from the point of view of the airport or air traffic.

☐ Only about 250 noise-related complaints are received each year, although Zürich Airport lies only a few kilometres from the city of Zürich, and the area surrounding the airport is densely populated. Zürich is one of the ten largest airports in Europe with 242,500 (227,900 IFR, 14,600 VFR) annual movements in 1994.

☐ Since 1985, Zürich Airport has had its own independent flight track monitoring system. It receives, without any formal agreement, the radar data on-line from ATC. Most European airports are not in this enviable position.

☐ A team of eight people has been formed to address these noise-related issues; although only one officer is responsible for all other environmental issues.

These facts illustrate that Zürich is more the exception than the rule in handling noise-related problems. To understand the unique issues it is necessary to outline some general aspects concerning the political system, and the consequences of the definition of noise, including empirical results.

THE POLITICAL SYSTEM OF SWITZERLAND AND NOISE CONFLICTS AROUND THE AIRPORT

Switzerland's model of democracy is a so-called direct democracy as opposed to a representative democracy. Between elections the people have the right to vote on public initiatives or referendums. This includes referendums on financial affairs. For example, in the Canton of Zürich, any credit for investments in excess of 20 million Swiss Francs must pass a public vote!

What has that got to do with airports? The international airports in Switzerland (Geneva and Zürich) belong to the State (to the Cantons of Geneva and Zürich respectively). This means in effect that the airports are dependent on the goodwill of the people.

Figure 1 shows the percentage of votes in favour of airport/civil aviation initiatives on 13 separate voting occasions.

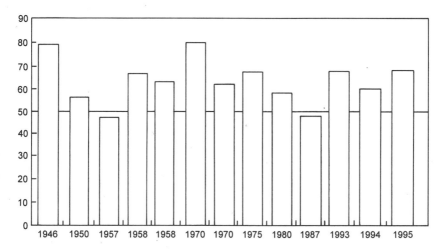

Fig. 1. Percentages 'pro airport/civil aviation' votes

The votes of 1946, 1950, 1957, 1958, 1970, 1975, 1980, 1987, and 1995 addressed financial issues, those of 1970, 1993 and 1994 legal or political questions. The 1995 vote concerned a credit of 873 mio SFr., being the Government's share of the fifth extension of Zürich Airport (Project Airport 2000).

The number of polls held indicates the extent of interdependence between aviation and the political system. It also shows that negative results for aviation are possible in reality, and that air traffic had and still has quite a high level of prestige. These circumstances have an influence on the action taken by the airport authority and the Government.

In every country (including Switzerland), the air transport industry has a very important position in the economy and has strong lobbying power in the political system. However, as Fig. 1 illustrates, this is no guarantee of positive results in public votes.

Despite experiencing economic difficulties (which is unusual for many Swiss people) environmentally conscious voters have, in the past, produced some astonishing results. For example, in 1994 we had a decision against the traditional politics concerning traffic through the Alps (with limits on the choice of traffic for goods transport), and in 1995, the people said no to the traditional concept of agriculture and yes to a greener production of agricultural goods. This is mentioned here as it illustrates that air traffic can not be sure of public support for expansion plans indefinitely.

A further rare element of the political system is the small size of political units. Switzerland, though a very small country, has 26 Cantons, which means 26 different and independent political systems. The whole country has more than 3000 communities which are more or less autonomous. The Canton of Zürich has 171 communities of which 18 lie inside the 45 NNI (Noise and Number Index) contours. The consequences for the airport and the Government is that they have to contend with at least 18 political conflicts.

To understand the behaviour of the organisations of air traffic in environmental and especially in noise affairs, it is important to reiterate and to understand what the definition of noise is:

> Noise is disliked sound.

Despite the simplicity of this definition, the problems associated with noise are complex. Recent socio-acoustical research,[1] based on interviews with 2050 people, showed that the correlation between Leq and annoyance is $r = 0.33$ and this means that 11% of the variance of annoyance can be explained by acoustics; most of the problems relating to noise affairs start with this correlation. This is nothing special to the experts and may therefore be considered a rule.

The low correlation explains why noise caused by air traffic has been such a longstanding problem. Contrary to a note from IATA to the airport authority in Zürich which stated 'The goals have been achieved' (concerning noise), further protests are to be expected. Although the overall number of protests may not increase, new and/or more complex issues may be raised. For example, in Zürich, there is a tendency towards a radicalisation in green questions among groups of the middle classes. This is of particular significance, as a large proportion of Swiss society is middle class.

Summary

☐ Annoyance seems individually so simple, but noise – especially the noise of aircraft – is a source of endless conflicts and an alibi for ideological fights, because the interdependencies between acoustics, physiology and socio-psychology are highly complex.

☐ This fact is independent of the position of individuals in the conflict between air traffic and society, in so much as each partner in the conflict is involved as in the above.

☐ The occurrence of noise problems and the means by which they are handled by the air traffic organisations depends on the structure of the organisation of power and/or the management of conflicts.

☐ In Switzerland, the degree of influence on decisions concerning the development of air traffic, the justification of decisions and the identification with the airport is high.

NOISE ABATEMENT AT ZÜRICH AIRPORT

Noise abatement at Zürich Airport must be analysed and described with due regard to the circumstances mentioned in the foregoing section and within the context of scientific, technological and acoustical history. For people outside the German-speaking countries it might seem amusing that the term 'abatement' in the German language is translated by 'Kampf', as being a military term denoting combat, fight, action, battle or struggle!

The approach is many fold and is described below.

The traditional approach: measuring noise

Since 1966, at Zürich Airport, we have been monitoring noise at nine monitoring sites in the near vicinity of the airport. Each month the results are published in a noise bulletin which is submitted to the local authorities, the residents committees, and politicians etc. The bulletin contains:

☐ number of movements for each runway and for each SID and STAR
☐ NNI and Leq for each measuring point
☐ detailed information concerning breaches in the night flight restrictions (exceptions like rescue flights etc).

Unlike at airports which have installed a noise monitoring system, there are no specified noise limits at Zürich. In the past we have defined such limits for four measuring points, but the more we worked with FANOMOS, the less we used noise limits. In reality only one type of aircraft, causes trouble: the B747-300. However, the problem is as much a function of the limits as of the plane!

In addition to the fixed monitoring sites the airport has a mobile measuring station. This is used to check the results of aircraft noise simulations and, unfortunately, for political reasons, to measure noise levels at particular localities. This is regarded as unfortunate because measuring noise does not in itself make a situation quieter but only serves to arouse conflicts.

Flight track monitoring: the optimal instrument

Since 1984, a flight track monitoring system which is currently being renewed, has been installed at Zürich Airport. Prior to this date, flight tracks were checked by reference to the noise monitoring stations which were situated adjacent to the flight paths.

People have an acute perception of changes, and with an optimal system of control we can guarantee that no unwanted changes will occur. Consequently, the most important use of the flight track monitoring system is to influence the behaviour of the pilots – to get them to fly the SIDs, which are defined as minimum noise routes as accurately as possible.

The method by which this is done is outlined below.

1. Zürich Airport receives radar data on-line from ATC.
2. Each IFR take-off is automatically analysed for accuracy.
3. For each SID, tolerance limits are defined. These limits are independent of PANSOPS criteria. Zürich Airport's tolerances are only defined on the basis of experience with the goal of the flight track control. Because the SIDs are defined as minimum noise routes, the goal is to draw the tracks as precisely as possible along the ideal or theoretical routes.
4. Each deviation from the defined path identified by the software is reviewed by airport staff. The review includes:

    – a voice check (radio communication between PIC-ATC)
    – weather conditions at the time of the incident.

    If the results indicate an inexplicable or inexcusable deviation, a letter is sent to the company in which the pilot in command is asked to submit a written statement within 30 days explaining the deviation. A radar-plot of the theoretical and the actual flight track (without tolerances mentioned) is enclosed.

5. If, after 30 days, the answer is outstanding, a reminder is sent with an additional deadline of 15 days.
6. After having received an answer from the pilot, the aircraft noise abatement department closes the case with a written confirmation

to the company after having evaluated the facts and the explanations given by the pilot.

This procedure, which is used in 99% of all cases, is principally aimed at increasing the awareness and altering the attitude of pilots (and not at imposing financial or legal sanctions).

At Zürich Airport, about 300 to 400 procedures are handled in this way each year. Approximately 50% of the resultant statements have a professional status, 30% of the pilots cannot remember the reasons, the rest are 'story tellers'.

7.   If the Airport Authority does not receive any reaction, the documents are sent to the Federal Office for Civil Aviation (FOCA) for further handling.

8.   If the deviations include serious mistakes (for example a left turn instead of a right turn), which have safety rather than noise implications, the documents are sent (see point 6) to FOCA for an official legal investigation, which may lead to actions against the pilot.

A side effect of this system is the increased contact between pilots and noise abatement staff. This is valuable for both parties as it ensures a continual learning process. Additionally, Swissair organise refresher courses for their pilots where airport staff are able to present the most common and critical experiences leading to deviations, and the special problems associated with noise abatement can be discussed. Similar courses are planned for the ATC. Finally, the airport also produces a monthly bulletin for the operators, in which the results of the flight track monitoring system are presented.

Complaints – ritual or political instruments?

The noise abatement department of the Airport Authority can be readily contacted at all times by telephone; the telephone number for complaints is in the directory and is also published in some newspapers. Each complaint receives a response, usually by telephone.

Complaints generally fall into one of two groups:

☐   well-known 'customers' with a political interest or, in a few cases, personal problems
☐   a group of occasional complainants; they react to special events like a late and/or unexpected departure.

As mentioned in the Introduction, the average number of complaints is about 250 per year, although in 1994, 350 complaints were registered. This situation is not comparable with other airports, because thanks to the organisation of the political system, the behaviour of the airport is endorsed by the decisions of the voters. This is also the reason there are few legal struggles between the airport and the neighbouring residents and/or the local authorities.

Complaints, especially those which are voiced in the media, or which use a political channel (via local political parties and/or representatives of those parties and which are sometimes directly referred to the minister) are unpopular. Despite the usual, official, negative feedback, these actions provoke internal tactical discussions because, of course, not every complaint, criticism etc. from outside may be ignored.

The complaints are frequently used as an instrument by middle-class suburban residents who are often exercising local political influence and/or have a hidden agenda (for example, qualification for tax concessions on housing. Furthermore, the real 'Greens' and/or 'Leftists' do not usually complain – the votes are their forum.

The number of complaints at Zürich Airport is therefore a bad indicator for judging disturbance due to the noise of aircraft.

Night flight restrictions

Since November 1972, Zürich Airport has had a system of night flight restrictions. The system was imposed as a condition for the third phase of airport expansion. Both the restrictions and the airport expansion had generally been ratified in referendums.

The system is quite complex and relies on the classification of traffic into three categories (Appendix A):

☐ scheduled traffic, with the highest priority
☐ charter flights
☐ general aviation, which has the lowest priority.

For general aviation the ban starts at 22:00 and ends at 06:00; the scheduled traffic is banned between 00:30 am and 05:00. So the effective ban is for 4.5 hours. The differentiation between scheduled traffic and charters was originally made because the charter operators generally used older and noisier aircraft. Today, this reason is largely obsolete and the federal office for civil aviation FOCA and pressure groups are trying to get rid of the differentiation.

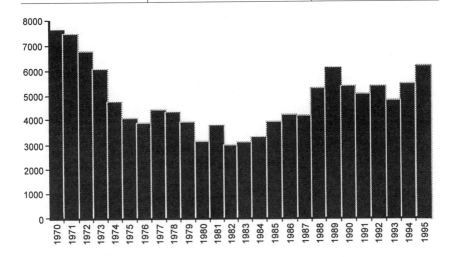

Fig. 2. Number of movements between 22:00 and 06:00

Figure 2 illustrates the effect of the restrictions between 1970 and 1995.

The Government of the Canton Zürich gave a clear guarantee, prior to the vote concerning the fifth phase of the airport expansion, that the airport would adhere to the system in use.

Noise-related surcharges

Noise-related surcharges have been in force at Zürich and Geneva Airports since 1 November 1980. The regulations are laid down in the Swiss Aviation Handbook (AIP) FAL 3-1 LSZH 1 Art. 5a and FAL 3-1 APP A.

International noise certificates according to ICAO Annex 16 or FAA FAR 36 will not be considered or accepted because of the principle 'the larger the aircraft, the greater the permissible noise'.

The charge imposed by Zürich Airport Authority, which is also responsible for controlling the noise situation around the airport, is based on the noise levels actually recorded during take-off at the three noise monitoring terminals in the residential areas near Zürich Airport.

Both Zürich and Geneva Airports (and the political authorities) are firmly convinced that this method of noise-dependent charging should continue to be used in coming years.

The average peak noise level of all departing jet aircraft is calculated for runways 16, 28 and 34, and the respective microphones. Similarly, the average peak noise level of each individual aircraft type is also calculated for each of the three runways and the difference between this and the average value for all aircraft is determined. The aircraft type can then be classified into one of five classes and the appropriate noise surcharge can be levied:

- □ for class 1   SFr. 800.-
- □ for class 2   SFr. 400.-
- □ for class 3   SFr. 200.-
- □ for class 4   SFr. 100.-
- □ for class 5   no surcharge is levied.

Experience suggests that a three-year period should be set during which the classification of the aircraft types would remain unchanged. Prior to the expiration of this three-year period, the system would be reviewed in the light of the preceding year's values and the classification would be modified in a formal amendment procedure.

The noise surcharge or noise-dependent landing charge at Swiss airports does not constitute a dissuasive levy in the true sense of the term.

Since its introduction, the existing noise surcharge or noise-dependent landing charge has become a separate airport charge, like the weight-related landing charge or the passenger charge.

The earnings from the noise surcharges are primarily used for covering the aircraft noise cost elements, such as the aircraft noise fund and noise abatement measures. But, because the laws of civil aviation do not foresee any systematic noise insulation programmes, very little money has been spent on these issues. In the future, when aircraft noise may be regulated by federal environmental laws, there may be a legal basis for realising noise-related insulation programmes.

Land use planning

Shortly after the start of the jet age at Zürich Airport, the federal government commissioned a task force to investigate the annoyance caused by the noise from air traffic. The principal aim was to define an official measure for aircraft noise and, based on these results, to produce a land use plan.

The effect of the noise of air traffic in terms of annoyance was investigated in a large socio-psychological and acoustical study carried out in 1971/72.[2] One result of the study was that the Noise and Number Index (NNI) became the official measure for noise of aircraft for the international airports at Zürich, Geneva and Basel. The legal basis for the land use plan was dated January 1974.[3]

In 1977, nearly ten years after the initiative of the Government, a specific land use plan for Zürich was published. The large number of protests led to a reassessment and in 1982 a revised version was published. This resulted in around 150 protests being registered, of which about 50 provoked long legal proceedings. In 1987, some 20 years after the project was initiated, the land use plan was legally endorsed.

In the intervening period, communities defined their own land use plans ignoring the problem of aircraft noise. The land became the object of speculation by property owners, investors and communities (communities equated growth with wealth). Consequently, by 1987, communities and individuals on the one side and the airport authorities on the other were in direct opposition with regard to land use planning. On the one hand, there was a demand for a quiet unspoilt environment for residential use, whilst on the other hand, there was a pressing need for expansion by the air traffic system.

The land use plan was designed only as an instrument to influence the use of geographical space and to reduce possible conflicts between different users or requirements. Protection of people from noise was not the priority aim of the plan.

The plan outlines three zones: Zone A which lies within the precincts of the airport, is reserved for buildings of the airport. Residential development is prohibited for obvious reasons. In Zone B (55 to 65 NNI) industrial buildings are permitted; and in Zone C, between 45 and 55 NNI, housing with special noise insulation is allowed. The cost of the insulation is borne by house owners/developers.

It is important that the concept of the land use plan is supported by federal laws for civil aviation. It may be a speciality of the Swiss solution that, despite the noise-related surcharges, the polluters need not pay the costs for noise insulation and that the regulations relate to new houses only. The legal basis of the law for civil aviation is due to be reviewed in the light of the laws for environmental protection. One of these legal foundations indicates that the polluter will have to pay without discrimination between old and new buildings. The protection of people is the basic aim of this law. We do not know when this change will come and how strict the limits will be. However, it is certain that Switzerland will change from NNI to Leq.

Conflict regulation

With regard to the variation of aircraft noise related annoyance, noise is only partially an acoustical question because of the heterogeneous interpretation of sound, especially sound of aircraft. Agreement in noise affairs is definitely impossible. That is why noise is always in conflict with various implications. Normally noise plays a role in all air-traffic related projects. A specialist in noise-related affairs once said that noise abatement staff should be fired at the latest when an airport is closed!

Furthermore, as the political situation is quite difficult, a strategy to handle the conflicts is required. Zürich Airport has no formulated concept, but some institutionalised or ad hoc instruments for conflict regulation.

Noise Commission

In the civil aviation legislation, valid until 1.1.95, noise commissions were prescribed. The Commission at Zürich acted as a consulting body for the Government in all noise affairs of the airport and, although it had no statutory powers, it had its own initiatives. In the new legislation – which saw a concentration of power with the Federal Office for Civil Aviation – this prescription has been withdrawn! Despite this, the Canton of Zürich will retain its Commission in the future on a voluntary basis. Politically, abandoning such commissions would be entirely the wrong approach to take. The following interests are represented within the Commission:

- [ ] Airport Authority   2 members
- [ ] Co-ordination office of Canton Zürich
        for environmental matters   1 member
- [ ] science (acoustics)   1 member
- [ ] Swissair   1 member
- [ ] Swisscontrol (ATC)   1 member
- [ ] association of residents   1 member
- [ ] Town Kloten   1 member
- [ ] Canton Aargau   1 member
- [ ] Germany   2 members.

The actual president is a representative of the Airport Authority. The Commission meets three times a year.

One of the most interesting projects which was realised in the last two years by the Commission, was a study on the issue of whether to spread or concentrate aircraft noise.

The project 'Spreading or concentrating aircraft noise'

A. Background and objective of the project   The question of distributing aircraft noise emissions caused by air traffic is a constant subject of public and political discussions. It acquires specific significance in flight-track changes or adjustments, as well as with other projects such as the increase of airport capacity, construction of new runways, etc.

A member of the parliament of Canton Zürich asked the Government to check the current system of departures and approaches with particular regard to the distribution of noise emissions and the effect on this of, for example, changing the altitude in an SID at which a left turn is initiated or amending the distribution of traffic between runways. Section B describes the scientific methods which were used to collect the empirical data for the Government's reply.

The evaluation of different measures concerning the effects of noise distribution, the study 'spreading/concentrating', should also be a help in developing guidelines for judging aircraft noise related projects and measures in the future (see Section D).

B. Situation of the noise effect research   There are several means by which the effect on noise of organisational or operational changes can be assessed. Conventional considerations are limited to purely acoustical facts. In other words, in case of certain measures (for example a new spread of traffic between two runways) purely acoustical differences (in terms of Leq, NNI, etc) are calculated. The quality of this approach is limited, as the relationship between noise levels and annoyance is normally explained with insufficient empirical and statistical justification. Zürich Airport's experience with the purely acoustical calculations and the estimations of their effects on the population has generally been bad.

Progress in acoustical and social research in the last few years shows a way to overcome the above mentioned limitations. Based on large acoustical and social surveys[1] concerning the relations between Leq

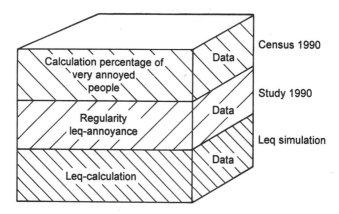

Fig. 3. The methodical link of basic empirical data

and annoyance, it is possible to describe this relationship in mathematical terms. In Switzerland the data of the 1990 census is available on the scale of hectares (100 m x 100 m). A relationship between the three factors: acoustics, annoyance caused by aircraft noise, and density of population can be established (as shown in Fig. 3).

For a number of reasons it makes sense to calculate the effects of organisational and/or operational measures for various geographical zones. This implies the possibility or the necessity of defining an overall-noise-system. The present basic data in Switzerland currently requires that the margin of this overall-noise-system is set at a 51 dB(A) curve for pure aircraft noise (Leq, day). For regions in the vicinity of the airport this limit is close to the daytime ground noise level.

Apart from reflecting on the whole, it is necessary and significant to analyse the changes in smaller zones. For this purpose an analysis is made in each zone; and a distinction between districts is possible (shown in Fig. 4).

The determination of annoyance is made in two stages. The percentage of people who feel strongly disturbed by aircraft noise serves as the basis to determine annoyance. The effects of the disturbance are elicited by a survey. Levels of 8, 9, or 10 on a scale of 11 points are interpreted as a strong disturbance. This data is correlated with the scale of acoustical data. The calculation of the annoyance is based on these relations. Annoyance expresses the percentage of a defined basic population (in total), which is strongly disturbed by aircraft noise.

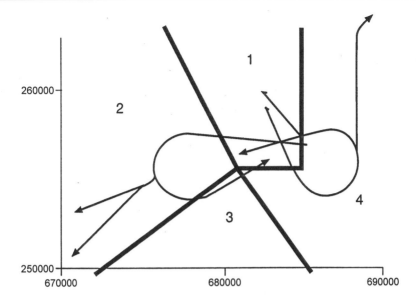

Fig. 4. Definition of zones 1–4 (arrows show SIDs)

C. Results In order to investigate the political requests, scenarios had to be defined (as described above).

Generally, once aircraft reach 5000 feet above sea level, the ATC can direct the planes off the prescribed SID and lead them by a left turn to the next navigation mark. With increasing performance, the point at which aircraft are able to initiate the turn has shifted nearer to the airport and this trend is likely to continue in future. This leads to a continually growing concentration of air traffic in the vicinity of the airport.

The exemplary single case. In Fig. 5 the effects on annoyance by raising the turning point when departing from RWY 28 (Departure Zürich East) are demonstrated.

When analysing the results it is important not only to judge the absolute, but also the percentage change of the number of people who experience a strong disturbance by aircraft noise. The reason for this is that the number of people of the present state is not the same as the number of people in the 'scenario after change'. Due to the 'measure 6000 ft', the basic number of people of the 'scenario of change' is 201,800 (compared to 209,300 people in the 'present state'). But the measure also provokes an increase in the number of people who

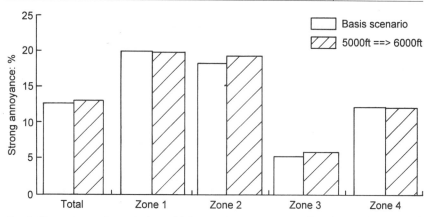

Fig. 5. Percentage of population which experiences a strong disturbance from aircraft noise – comparison between the present state and the scenario of change from 5000–6000 ft

experience a strong disturbance from 12.6% to 13.2%. This increase is explained by the new aircraft noise situation in the Zones 2 and 3. In these zones the surface of the same Leq has increased.

In Zone 1 the simulation creates no change. This is an indication of how plausibly the model reacts, as in Zone 1, the first segment is the area which is influenced by the initial climb and therefore remains unaffected by the measure.

Summary of the main results. The analysis of the five calculated scenarios produces an important main result. Each calculated single measure leads – with one insignificant exception – to an increase in the number of people who are strongly disturbed by aircraft noise. This is, in other words, a strong (but not conclusive) indication that the distribution of aircraft noise in the vicinity of Zürich Airport is at its optimum.

This result indicates that the question of spreading or concentrating aircraft noise is not correct. The correct question is: Is the spreading of annoyance at its optimum? In this case 'optimum' is defined as those circumstances in which aircraft movements, SIDs, organisational concepts (time!) accommodate the structure of population, so that each measure, in the sense of a single measure, leads to an increase of annoyance.

D. Guideline for judging aircraft noise related projects and questions It has been mentioned that the analysis should serve to develop guidelines and, based on that, provide recommendations for judging aircraft noise related projects. In the opinion of the experts in the special task force of the commission (the residents were represented in this group of five) the guidelines should include the following questions.

☐ Are the decisions based on the present state of knowledge or on experience? Have acoustical effects and the effects on annoyance been investigated?
☐ Does a measure imply relevant noise related consequences?
☐ Has a simulated state of distribution of noise been compared with a real situation in the year before?
☐ What happens if a measure leads to more annoyance?

The Commission also formulated a number of recommendations to assist in responding to each question. These recommendations are addressed to those who are involved in decision-making, including the minister in charge. The future will show whether these guidelines have an influenc – in the sense of a higher objectivity of decisions.

Project-orientated task forces

In the past, the air traffic authorities tended to establish small project teams which tended to put forward proposals without offering any alternatives. Furthermore, other interested parties such as residents associations and the noise commission were consulted at a late stage in the project development. However, there has been a gradual increase in the number of task forces being established which include representatives of all the concerned parties – including those from the local community. The result was the danger of 'projects with a one sided approach', misunderstandings, frustration etc.

Reaction to the use of these multi-representative task forces has been quite positive. Both sides have had to learn how to prepare basic material for decisions, to learn to accept (sometimes unpopular) facts and to construct their arguments using facts.

Regular meetings with local authorities

A similar approach is to have regular meetings with local authorities on different levels of the airport authority's hierarchy. The meetings are intended to provide direct information on what happens at the airport in general and/or concerning special subjects. To use an example from the noise abatement business: any extraordinary results of measurements outside the fixed monitoring system are documented in a special report. The report is presented to the local authorities by the staff of the noise abatement department who in response to community demand make public presentations and/or help to formulate accurate information (but not interpretation).

Education

One of the more unusual measures is the noise abatement department's involvement in the professional education of Swissair and Swisscontrol (ATC). This can also be regarded as an instrument for conflict regulation. The intention of the courses is, on the one hand, to make pilots and controllers more aware of noise issues in the hope that they will reflect this knowledge in their work. On the other hand, these people are often consulted by their neighbours and others in noise affairs as well. There is often a tendency to act and react out of ignorance. These courses serve to make people aware of other points of view and to use their prestige as employees of the national airline in public discussions. A pilot who says, 'I understand your problems. The situation in the cockpit is as follows..., and we use such and such a procedure to reduce the noise' is doing a great deal to bridge the gap of distrust that exists.

RESULTS AND OUTLOOK

The means of reducing noise are, in reality, very small. If there is progress, the reasons have, to a great extent, no connection with the activities of noise abatement. That is the rule.

It is also the rule that, in general, economic and political reasons influence the technical progress and the decisions of the operators in noise affairs. Unfortunately, but naturally, technical progress concerning noise is near its limits. A technical revolution in this field is a long way off. The moment comes, at the latest after the real phasing out of Chapter 2 aircraft, in which the number of flights increasingly influences the size of the noise contours and increases the contours, for instance. The increasing traffic will also influence the operational times. Because the day is only 24 hours long and because the air traffic system has some problems with the capacity of runways, management of the increasing traffic in the air, etc., there will be pressure in the sense of more movements during the night time in the future.

So, in future, fewer technical and acoustical questions will be debatable. Increasingly, instruments to regulate conflicts will become important.

The presented analysis and description of some elements of the noise abatement activities in Zürich illustrate some original ideas and experiences (e.g. the model for the surcharges, night flight restrictions, flight track monitoring). But the situation is only comfortable as long as the promise of technical progress remains available.

It seems likely that the political system will not be able to guarantee easy further development of the noise debate around Zürich Airport. In addition to the traditional regulations of political conflicts it needs to create and test new instruments in the sense of 'rules in the game'. Some key points follow.

☐ Introduction and testing of possibilities of mediation in the sense of open planning processes.
☐ Abandoning of monopolies on information and making information readily available to all.
☐ Language has to be used for communication, not for domination.
☐ Guarantees must be made to the partners in a conflict with the goal of building ties of trust.
☐ Agreements must be controlled by the partners; no statistical tricks.
☐ Consensus must be achieved concerning facts, but not concerning interpretation.

We are on the road to such action in Zürich, not in the sense of a decision of the Board, but more as an intuitive process.

So, returning to the idea of 'the exception and the rule': in the past Zürich was more an exception than it is today. In the future there will be a tendency towards a decrease of effective progress in noise abatement and thus a lessening of its influence in debates. Unfortunately, this seems to be the rule.

REFERENCES

1. Lärmstudie 90: Belastung and Betroffenheit der Wohnbevölkerung durch flug - und Strassenlärm in der Umgebung der internationalen Flughäfen der Schweiz, C Oliva (1993).
2. Sozio-psychologische Fluglärmuntersuchung in Gebiet der drei Schweizer Flughäfen Zürich, Genf und Basel, Bern, 1974.
3. Verordnung über die Lärmzonen der Flüghafen Basel - Mühlhausen. Genf - Cointrin und Zürich, 23 November 1973.

APPENDIX A: REGULATIONS ON TRAFFIC RESTRICTIONS AT NIGHT

(NIGHT FLIGHT PROHIBITION)

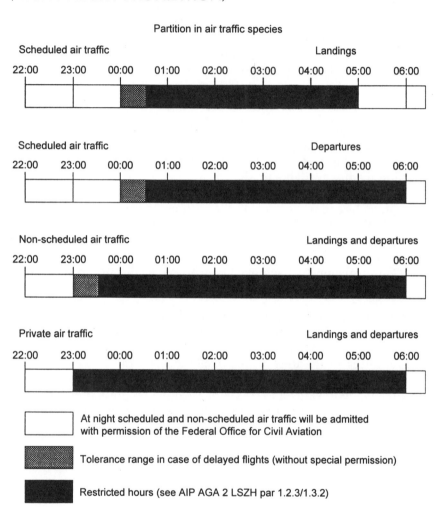

Partition in air traffic species

At night scheduled and non-scheduled air traffic will be admitted with permission of the Federal Office for Civil Aviation

Tolerance range in case of delayed flights (without special permission)

Restricted hours (see AIP AGA 2 LSZH par 1.2.3/1.3.2)

Changes in air quality processes in the United States

ROGER L WAYSON, University of Central Florida, Civil and Environmental Engineering, PO Box 162450, Orlando, Florida 32816-2450, USA

INTRODUCTION

Air quality methodologies and computer tools at airports in the United States have been developed and used since the mid 1970s. However, these procedures and tools have had to be dynamic as changes and improvements (primarily minor) continued to occur. Within the last three years, the changes have been much more pronounced, leading to several large changes in procedures and models. These changes have included promulgation of various versions of the computer model EDMS, changes required by the 1990 Clean Air Act Amendments such as conformity, and an emphasis toward more use of the private sector instead of relying on employees of the US Federal Aviation Administration. This paper describes many of these changes and discusses the impact on air quality practices. Special emphasis is placed on those changes directly affecting the air quality modelling process.

CONFORMITY vs NEPA REQUIREMENTS

Before 1970, air quality analyses were not regularly undertaken for any US airport project. On 1 January 1970, the National Environmental Policy Act (NEPA) was signed into law and required an Environmental Impact Statement (EIS) to document possible impacts from any major Federal action in the United States. Since 1970, an Environmental Assessment (EA) or EIS has become standard practice for any large airport project since funding often comes from the US Federal Aviation Administration (FAA). The NEPA analysis, whether an EA or EIS, requires future scenarios to be compared to the existing situation and to each other. The objective was to determine alternatives that would result in the least environmental impact rather than justifying a planned project. To this end, analysis of the 'do-nothing' alternative is always a requirement. In addition, comparison to the National

Table 1. National Ambient Air Quality Standards (NAAQS)

Pollutant (units)	Averaging Times					
	1-hour	3-hour	8-hour	24-hour	1-quarter	1-year
Carbon monoxide (ppm)	35		9			
Nitrogen dioxide (ppm)						0.05
Ozone (ppm)	0.12					
Suspended particulate:						
PM$_{10}$ (μg/m^3)				150		50
				(150)*		(50)*
Sulphur dioxide (ppm)		(0.05)*		0.14		0.03
Lead (μg/m^3)					1.5	

* Values represent secondary standards

Ambient Air Quality Standards (NAAQS) is used as an absolute standard. The NAAQS are shown in Table 1.

The NEPA analysis at airports is approached somewhat differently for airports than for other mobile sources. Fig. 1 contains a flow chart of the process normally followed at airports.[1] If airport operations are not changed, assessment of air quality may not be required. In addition, smaller airports are not required to perform this analysis as shown in Fig. 1. The process usually begins by identifying all major sources directly related to the airport operations and quantifying the total emissions for each alternative by use of an emission inventory. Other mobile sources usually do not include sources such as fuelling operations when evaluating a project. However, at airports, all air pollution sources under airport control must be evaluated. As shown by Fig. 1 and previously mentioned, an emission inventory is the primary tool in the first phase of analysis. Usually included in this inventory is carbon monoxide (CO), oxides of nitrogen (NO$_x$), and hydrocarbons (HC; especially volatile organic compounds, VOC).[2] In some cases, such as the Pittsburgh International Airport, an emission inventory for suspended particulate matter less than ten micrometers in aerodynamic diameter (PM$_{10}$) may also be required. However, dispersion modelling is not required unless two conditions occur:

☐ there is non-conformance with the State Implementation Plan (SIP) and/or
☐ there is potential for the CO standard to be exceeded.

If either of these conditions occur, dispersion analysis must be accomplished for carbon monoxide, a conservative pollutant. The Airport and Airway Improvement Act of 1982 required that prior to

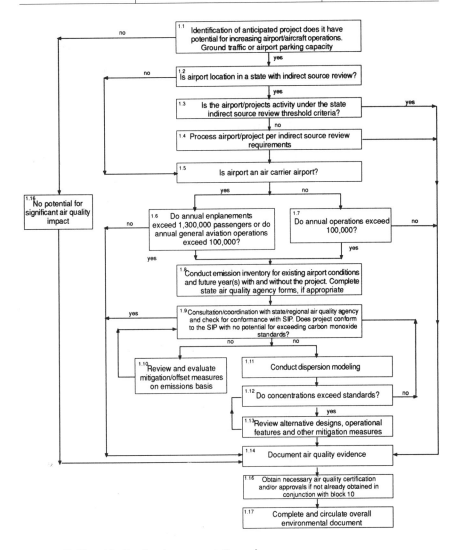

Fig. 1. Civilian Air Quality Assessment Procedures

implementation of a major airport improvement project, the FAA must receive written assurance that the NAAQS will not be exceeded.

It is important to understand that, in the United States, areas that exceed any NAAQS have been listed as Non-Attainment Areas (NAAs) since the 1977 Clean Air Act Amendments. In these NAAs, source controls are required. At present, the most prevalent NAA problem is for the secondary pollutant ozone. Particulate matter and carbon monoxide are distant seconds. In states that contain NAAs, a State Implementation Plan (SIP) must be filed with the EPA. The SIP acts as a contract between the states and EPA on exactly how the state will come into compliance. To ensure that SIP requirements are met, projects proposed in the NAA must show conformity to the SIP, as shown in Fig. 1. The 1990 Clean Air Act Amendments strengthened this requirement for conformity. The 1990 Act also listed areas by severity, each with different times for compliance and different source reduction requirements.

In 1993, regulations were released as 40CFR51 and 93, the Final Conformity Rules. These rules required that other pollutants with regional impacts be included in the air quality analysis for NAAs. Since regional analysis using tools such as the Urban Airshed Model are beyond the scope of most projects, this regional analysis has centred around emission inventories. If it can be shown that the preferred alternative results in a reduction of NO_x, VOCs, CO, and/or particulate matter less than 10 micrometers in aerodynamic diameter, (PM_{10}), the project 'conforms' to the SIP and is 'in conformity'. If the project is determined not to be in conformity, it can be stopped as has happened when NEPA analysis showed that NAAQS were exceeded.

In addition to the conformity determination, NAAs also have 'caps' established for overall area emissions. Initially established in 1990, the 'caps' are updated on a regular basis, usually three years. For a defined year, emissions of a specific pollutant such as NO_x cannot be increased from this baseline condition. This led to difficulties since the airport emissions may or may not have been considered in the emission inventory that 'set the cap'.

Since the emission inventory is usually the first phase during airport air quality analysis, conformity analysis requirements required less changes in methodology than those required for other mobile source analysis where emission inventories were not generally required. However, last minute conformity delays are continuing to occur, such as at the new Denver International Airport where the project experienced a three month setback due to conformity determination considerations.

In an effort to identify and correct such delays, the FAA held EIS Air Quality Workshops in Seattle, Washington; Atlanta, Georgia; and Research Triangle Park, North Carolina between May 1992, and August 1993. Early interagency coordination was identified as the primary cause of conformity delays. To avoid this problem as well as others, the FAA has begun to develop flow charts in an effort to simplify conformity events. One such effort is shown in Fig. 2.[3] Fig. 2 should be used as appropriate with Fig. 1.

The flow chart[3] was presented at the 1994 Air and Waste Management Association Annual Meeting but is expected to be incorporated into applicable FAA Orders such as 5050.4A.[4]

It should be noted from Fig. 2 that the process requires determination of project type and overall impact on the Air Quality Management District (AQMD). The project is determined to be in conformance with the SIP if:

☐ emissions are reduced
☐ emissions are small when compared to the AQMD emissions
☐ the cap is not exceeded
☐ the NAAQS are not determined to be violated and/or
☐ the project is included in the local Transportation Improvement Plan (TIP).

It is easy to understand how delays might occur and how the process might be confusing.

Air quality analysts for airport projects must now consider both the NEPA and the conformity process or the project will experience delays or even be halted.

MODELLING

Since 1983 the preferred guideline model promulgated by the FAA and certified by the EPA has been the Emission and Dispersion Modelling System (EDMS).[5] This model has gone through many changes and has evolved into a better model with each change. However, additional changes are still required. These changes range from basic reconfiguration, such as including a greater number of aircraft types in the emission database, to reprogramming such as the addition of taxiway emission considerations. However, the individual that has been in charge of the model development is no longer with the FAA. Realising the needs for updating the model, FAA has recently passed along development to others and has stated that the model alterations will include:

☐ management of data through a standard database (.dxf files)
☐ operation in a Microsoft Windows environment

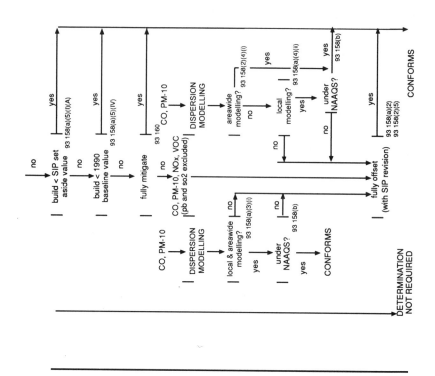

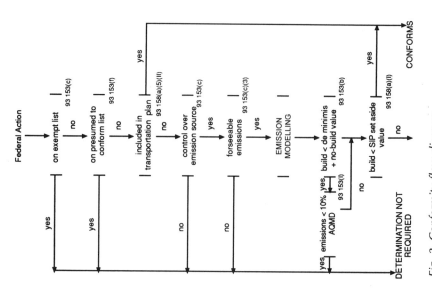

Fig. 2. Conformity flow diagram

☐ design for a 486 platform and
☐ compatibility with the FAA Federal Aircraft Engine Emissions Database (FAEED).

The expected completion date for the new model is expected in about twelve months. These changes will result in a more user friendly model (visual programming is being considered), more flexibility (the user must now determine which type of aircraft are similar for emissions), faster performance (the model is somewhat slow at present), and allow engine updates as they occur in FAEED. This last change is quite significant. At present, emission rates are known for certain engine types, especially newer engines, and are included in the FAEED. Unfortunately, reprogramming EDMS is not easy and these emission rates have not been included. The user may manually input emission factors, but this is quite time consuming, requires an experienced user, and may lead to mistakes.

It goes without saying that the new model is being anxiously awaited.

ODOUR RESEARCH

A common complaint at airports is odours. Research at Battelle Labs in 1984 identified sixty-four different species of hydrocarbons in the CFM-56 engine exhaust at idle using JP-4 fuel.[6] The CFM-56 engine is commonly used on aircraft such as Boeing 737 and Airbus A320/A321/A340. The dominant hydrocarbons were ethylene, formaldehyde, methane, propylene, and acetylene. Odour threshold concentrations were found for sixteen of the sixty-four hydrocarbon species using a 50% odour recognition threshold concentration. Similar results were reported for other engine types. From this example, it is easy to see the problems facing the air quality analyst or the individual dealing with odour complaints. What quantities exist of each hydrocarbon species and which are the real odour problems? This problem is compounded when it is considered that new commercial aircraft must meet EPA/ICAO standards for total hydrocarbons only. Relative percentages of individual hydrocarbons remain in question. Further complications arise when dispersion must be considered at the airport.

Logan Airport in Boston was dealing with such problems. In 1992 it was decided to proceed with a study to identify odorous compounds at the airport. Under the charge of Mr Norman Faramelli, Tech Environmental Inc conducted an extensive survey.[7] Seventeen species of hydrocarbons were measured at locations around the airport, using evacuated canisters as outlined in EPA method TO-14. Of these, three suspected odorous species were measured at a concentration exceeding the 50% odour recognition threshold concentration; acetaldehyde, formaldehyde, and naphthalene. Benzaldehyde was also suspected as a contributor to the odour complaints based on the measurements. Odour complaints, aircraft operation, and prevailing winds were also correlated.

While the Logan study does not completely answer all questions, it does help to show that jet exhaust is most likely to be the prevalent odour problem. More research is needed to identify the relative amounts of each species.

INDIRECT CHANGES

Noise regulations

In 1990, the FAA released strong noise control regulations as 14CFR91 (Transition to An All Stage III Fleet Operating in the 48 Contiguous States and District of Columbia) and 14CFR161 (Notice and Approval of Airport Noise and Access Restriction). These regulations have changed the general fleet that would have been expected in the near future and changes the distances that might be expected to receptors.

Of most importance is the change to a Stage III fleet. This change will not only make airports quieter, but will result in a significant decrease in aircraft CO and VOC emissions as airlines are forced to modernise their fleets faster than would have normally occurred. The regulations require a change to an all Stage III just after the turn of the century. This rapid change will result in newer engines which are more efficient and have reduced emissions of CO and VOCs. However, NO_x will continue to be a problem because, while the newer engines reduce the CO and VOC emissions, NO_x emissions are increased.

Programme continuity

The FAA air quality programme has been quite stable in past years because of a continuation of personnel. However, recently, key personnel have left to explore other careers. The result will be a lag period as new personnel continue to gain experience. There is now a small problem at FAA relating to responding to air quality questions. FAA has also allowed more private sector involvement than in the past for critical decisions because of needed personnel. For example, the private sector is sometimes engaged as the reviewer for a specific airport project. This strategy has both advantages and disadvantages. Experience lost at FAA will also be recovered with time.

THE FUTURE

Interest continues in several key areas relating to air quality near airports. These areas of interest include alternative fuels, odour control, impacts on stratospheric ozone, reduction of 'hot spots' caused primarily by motor vehicle traffic, and strategies for controlling NO_x.

Alternative fuels for ground support equipment would be easy to accomplish and would result in significant reductions of air pollutant emissions. US airports, particularly in California, are now exploring this option. Odour control continues to be a source of complaint. The rapid phase-in of Stage III aircraft should help to reduce this problem as total emissions of hydrocarbons are reduced. The newer aircraft should also help to reduce impacts on stratospheric ozone due to reduced emissions. Eliminating long idle periods for motor vehicles accessing the airport and increased use of fringe parking should help to eliminate exceedance of the NAAQS on the ground side of the terminal building. High speed taxiways and future engine research should help to reduce NO_x emissions. Combined, all of these options will help to significantly reduce air quality problems around airports.

CONCLUSIONS

Conformity analysis has become as important as the NEPA analysis. Air quality analysts in the USA must continue to develop this expertise. Further guidance from FAA, combined with experience when dealing with the somewhat confusing regulations, should lead to a comprehensive plan for air quality at airports. New releases of documents such as FAA Order 5050.4a will be essential to future improvements.

The computer dispersion model EDMS has served well but still has areas that need to be improved. The new FAA effort should result in a more user-friendly model, with greater flexibility, and better performance. This new model should be released during the summer of 1996 if the FAA schedule is maintained.

Odours continue to be a problem, but newer aircraft should help to reduce this problem since hydrocarbon emissions will be reduced.

Research, such as that at Logan Airport, will continue to help identify the specific problem, and should help to form a solution.

The rapid phase-in of Stage III aircraft will also help to significantly reduce emissions of CO and VOCs with the newer, cleaner engines. NO_x may continue to be a problem in the short term.

Changes in personnel at FAA will cause problems in the short term, but this problem is inevitable. Experience is hard to replace. Use of the private sector will help this situation but has both advantages and disadvantages.

Future changes in airport operations and aircraft should continue to reduce air pollution near airports but more research is still needed. We will all be winners if air quality is improved.

REFERENCES

1. Eberle, G.F. and Steer M.D., *Air Quality Procedures for Civilian Airports and Air Force Bases*, Report No FAA-EE-82-21, US Department of Transportation, Federal Aviation Administration, Washington DC, November 1982.
2. Wayson, R.L. and Bowlby W., Inventorying Airport Air Pollutant Emissions, *J. Transp. Engng*, American Society of Civil Engineers, Vol 114, No 1, pp 120, January 1988.
3. Segal, H.M., *Problems in Airport Conformity Determinations and Steps That Have Been Taken to Solve Them*, Paper No 94WP99.01, Air and Waste Management Association 87th Annual Meeting, Cincinnati, OH, June 1994
4. Federal Aviation Administration, *Airport Environmental Handbook*, FAA Order 5050.4A, US Department of Transportation, Federal Aviation Administration, Washington DC, October 1985.
5. Segal, H.M., Kemp J.K., Hamilton P.L., *A Microcomputer Pollution Model for Civilian Airports and Air Force Bases User's Guide*, FAA Report No FAA-EE-85-4, US Department of Transportation, Federal Aviation Administration, Washington DC, December 1985.
6. Spicer, C.W., Holdren M.W., Lyon T.F. and Riggin R.M., *Composition and Photochemical Reactivity of Turbine Engine Exhaust*, ESL-TR-84-28, Batelle Columbus Laboratories, pp 39-43, Columbus, OH, 1984.
7. Tech Environmental Inc, Identification of Odorous Compounds from Jet Engine Exhaust at Boston's Logan Airport, Massachusetts Port Authority Technical Report, Boston, MA, December 1982.

Review of ambient air quality at major Canadian airports

C. LEONARD TAYLOR, Tradewind Scientific Ltd, Box 3262 Postal Station D, Ottawa, Ontario, Canada K1P 6H8

Ambient air quality monitoring has been conducted at major Canadian Airports for more than fifteen years. The primary objectives of this monitoring programme are to collect data on the concentrations of several principal gaseous and particulate air pollutant species (CO, NO, NO_2, NO_x, O_3, THC, TSP, and PM_{10}), to evaluate the recorded air pollution measurements with respect to National Air Quality Objectives as specified in the Canadian Environmental Protection Act and finally to assess the impact of airport and aircraft activities on the local ambient air quality.

Most of the monitoring studies have been carried out using a mobile air quality monitoring laboratory operated by the Airports Group of Transport Canada. This laboratory has been asssigned, on a rotating basis, to conduct three twelve-month air quality surveys at the major international airports since 1980. Instrument replacements and equipment updates have kept the system compatible with current conventional air monitoring technology. More recently, some airport sites have been evaluating and installing fixed air quality monitoring systems based on DOAS (differential optical absorption spectrometry) technology, capable of measuring a wider range of substances.

An analysis of all the recorded airport air quality data indicates that present activities at major sites such as Montreal-Dorval, Ottawa, Toronto, Calgary and Vancouver do have a measurable impact on the local ambient air quality, both on site and in the immediate vicinity. The principal influences observed are from CO and NO_x emissions by vehicles and aircraft in the main terminal areas and from particulate emissions originating from construction and maintenance activities. However, most of the measured airport air pollutant concentrations are well below the applicable National Air Quality Objectives.

While data indicates that current airport air quality compares favourably with urban air quality in Canadian metropolitan areas, the long term trend in urban areas is normally one of stable or decreasing average concentrations of carbon monoxide, oxides of nitrogen and particulates. It is therefore probable that increasing aircraft operations and associated traffic at the busiest will eventually result in airport air quality becoming inferior to that of the surrounding metropolitan regions unless mitigative measures are implemented.

INTRODUCTION

As part of a continuing commitment to ensure that national environmental objectives are being met at major federal facilities, Transport Canada has been conducting a series of three, nine and twelve month ambient air monitoring studies at Canadian international airports.

The studies are used to measure, on a continuous basis, carbon monoxide (CO), nitric oxide (NO), nitrogen dioxide (NO_2), and ozone (O_3). Total suspended particulate (TSP) is measured based on a twenty-four hour sampling period that is normally conducted twice per week. Recently total hydrocarbons and PM_{10} sampling capabilities have been added to the air monitoring program.

A review of the air quality data gathered during these projects indicates that the maximum Federal Acceptable Air Quality Standards were sometimes exceeded for all monitored pollutants, with the exception of carbon monoxide, which occasionally exceeded the Federal Desirable Air Quality Standards.

The general direction in ambient air quality standards is in maintaining the status quo or in making them more restrictive; little relaxation of standards is foreseen. The implication for airports is that it will become more and more difficult to meet air quality standards. In comparing Canadian National Air Quality Objectives with the World Health Organization (WHO) Guidelines, Canadian standards are for the most part the same, with the exception of carbon monoxide levels. However, the United States National Ambient Air Quality Standards (NAAQS) were significantly higher than both Canadian and WHO standards.

The studies are conducted using a Mobile Air Quality Monitoring System which was originally designed and constructed in 1978. It had been kept up to date over the last decade with continuous revisions, upgrades and additions to the instrumentation and equipment. In 1993, a new mobile air monitoring truck was constructed to replace the outdated truck. To ensure that monitoring continues, it is recommended that new equipment and instrumentation be added to the truck.

A review of the airport monitoring methodology employed to date indicates that for studies of this nature to be valid, in terms of relating actual air quality to the national objectives, monitoring periods of at least twelve months are required. The projects should be planned to encompass all seasons and thus all conditions. Since the air monitoring truck can only monitor one airport site a year, it is recommended that the major airports move to a fixed monitoring system such as that based on Differential Optical Absorption Spectrometry (DOAS); which is a complete on-line monitoring system that gives results twenty-four hours a day.

Previous airport studies and US airport experiences have shown that relatively high levels of pollutants are encountered around main terminal apron areas. It is recommended that air quality studies maintain their concentration around terminal area sites while continuing to use monitoring locations from the previous projects. This practice would be particularly useful because, in the terminal apron area, specific mitigation procedures are possible to reduce emissions (i.e. aircraft towing, use of electric service vehicles, etc.). Routine exposure of the travelling public and airport staff to high pollutant concentrations is also more probable in these areas.

Motor vehicle traffic is the general offender on the landside for pollutants such as carbon monoxide and nitric oxide and is exacerbated by aircraft, service vehicles, and stationary sources. However, airport employees can be affected on the airside of the terminal because of the close proximity to aircraft in the idle/taxi mode and service vehicles.

Air quality modelling, a technique used to predict what will happen to ambient air quality when contaminants are released into the atmosphere, has been employed since the 1970s in airport related projects, primarily in the USA. In a smaller number of cases, dispersion modelling is required. A review of the Emissions Dispersion Modelling System demonstrated that it is primarily useful for screening purposes and environmental impact assessments. Short-term predicted concentrations are often in error by factors of 2 to 10 and the modelling calculations cannot be used as a substitute for actual air quality monitoring. EDMS was used for the preparation of environmental impact statements (EIS) at Vancouver and Lester B. Pearson International airports.

The investigation included:

☐ a thorough analysis of the air quality data gathered at ten Canadian airports during fifteen separate monitoring projects;

☐ a review of the interviews and correspondence with officials and scientists from Provincial, National and International environmental and transportation organizations originally conducted for the 1991 report, and;

☐ an extensive search of the technical literature concerning airport air quality studies as well as the evaluation and testing of the Emissions and Dispersion Modelling System (EDMS).

Concern regarding airports and their potential impact on ambient air quality began with the advent of substantial commercial turbojet air traffic in the early 1970s. These concerns prompted the United States Environmental Protection Agency (USEPA) to set specific aircraft engine emission specifications by 1973. Substantial revisions have been made to these standards since that time.[2] The International Civil Aviation Organization (ICAO) formed a special committee on which Canadian officials actively participated.[3] The ICAO committee's final recommendations for addressing low-altitude emission problems included the establishment of pollution monitoring devices at active airports. It was further decided to select and develop suitable air quality models and set engine emission reduction standards, particularly for hydrocarbons.

With the creation of the new 300 generation jet engine, manufacturers have been able to reduce NO_x emissions considerably. Further, over the course of the latest recession, traffic levels have decreased by 7.8% since 1989, leading in turn to a decrease in air emissions at airports. However, given an expected return to pre-recession air traffic volumes, airport air pollutant emissions remain a source of concern due to the larger aircraft in use and a resultant rise in ground related activities.

In order to assess the impact of airport activity on air quality, Transport Canada has undertaken a long-term monitoring program at the major federal airports. It should be noted that these studies only address the issue of ambient air quality at airports.

It should also be noted that Dorval, Mirabel, Calgary, Edmonton and Vancouver International Airports, which are now operated by Local Airport Authorities, were under Transport Canada operation when the air monitoring studies were first conducted.

AIR QUALITY STUDIES

The 1991 ambient air quality report concluded that aircraft and aircraft-related activities produce only a small percentage of the total air pollutant emissions on a national basis. However, given that these emissions are concentrated over relatively restricted areas, airport pollutants have a significant impact on local ambient air quality. In fact, the overall emission density of the principal combustion-related pollutants on an active commercial airport is comparable to that of a typical urbanized zone.

The following are the major air pollutants of concern on and near airports:

Carbon monoxide (CO)   This is a product originating from the incomplete combustion of hydrocarbon fuels; in idling aircraft engines, ground vehicle operation, heating plants, etc. The Federal Acceptable Air Quality Objectives are 35 mg/m^3 for 1-hour and 15 mg/m^3 for 8-hour periods, as measured by an NDIR spectrometer.

Nitrogen oxides (NO$_x$)   Products of the high-temperature combination of nitrogen and oxygen (primarily NO and NO$_2$); in aircraft engines and other internal combustion sources. The Federal Acceptable Objectives for NO$_2$; 400 μg/m^3 for 1-hour and 200 μg/m^3 for 24-hour periods, as measured by a dual-channel chemiluminescent spectrometer.

Total suspended particulate matter (TSP)   These are composed of liquid and solid particles in the 0.1–100 μm range, from varied sources such as ash, soot, smoke, fumes and dust from combustion and erosion processes. The Federal Acceptable Air Quality Objectives are 120 μg/m^3 for 24 hours and 70 μg/m^3 for annual geometric mean, as measured by a TSP/PM$_{10}$ critical flow hi-volume sampler, capturing particles of 10 μm or less (sometimes referred to as respirable particles).

Hydrocarbons (total and non-methane hydrocarbons: THC and NMHC) These include a wide range of pure and impure hydrocarbons, such as methane, alkenes, aldehydes, ketones and terpenes. They come from fuelling activities and incomplete combustion processes. At present no Canadian Federal Objectives exist. The monitoring instrument in Transport Canada's system is based on Flame Ionization Detection.

Ozone (O$_3$)   This is predominantly a secondary product resulting from photochemical reactions, which is known to play an important role in the chemistry of NO$_x$ and HC. The Federal Acceptable Objectives are 160, 50 and 30 μg/m^3 for averaging periods of 1 hour, 24 hours and 1 year, respectively, as measured by a UV absorption spectrometer.

Sulphur dioxide (SO_2), normally a major pollutant constituent of many combustion processes, is not considered a specific problem at airports because of the low-sulphur fuels consumed by aircraft engines and the absence of large industrial point-sources. Thus, although Federal Air Quality Objectives exist for this pollutant, SO_2 measurement capability is not considered necessary for the Transport Canada System.

Volatile organic compounds, which are directly-emitted components of combustion processes, have gained notoriety recently since several compounds are considered carcinogenic and chronic exposure could cause health problems. However, criteria would have to be considered on a compound by compound basis. The monitoring of these compounds is a complex undertaking. Recently developments in DOAS (differential optical absorption spectrometry) type monitoring has allowed for an on-line measurement of certain organic compounds. At present, an on-going EPA project at Chicago's O'Hare Airport in the US is monitoring toxic organic compounds.

Hydrocarbons from fuel handling operations and idling aircraft engines may be present in relatively high concentrations in areas adjacent to airports. Given the abundant levels of hydrocarbons and organics emitted by airport related sources, a lack of specific hydro-carbon information may raise more questions. Therefore, consideration should be given to sampling for specific carbon-based compounds to both qualitatively and quantitatively establish hydrocarbon levels.

The Transport Canada Airport Air Quality Monitoring System is well equipped to measure most of the major air pollutants associated with airport operations. A program of continuous maintenance and upgrading of equipment has allowed for Transport Canada to conduct long term monitoring projects with accuracy and reliability.

The fifteen Transport Canada Airport Air Monitoring studies completed over the last fifteen years have been conducted with data acquisition periods of three and nine months. However, since March (1990) studies have been conducted over a twelve-month period; (Dorval 1992) (Vancouver 1990/91). Evidence from these and other USA and UK studies shows that with the large number of emission and meteorologic variables affecting ambient air pollutant concentrations, monitoring periods of at least six months are required to adequately determine an airport's air quality measured against the applicable short-term governmental standards. Long data recording periods are also necessary to permit the identification of major pollutant sources by the correlation of concentrations with wind-speed and wind-direction.

The selection of the monitoring period is likewise critical in terms of collecting representative data. Local climatological patterns may define specific conditions which enhance pollutant buildup. These periods of the year in which the conditions occur should be included in the selection of the monitoring period. For example, on days with low wind velocities and convective mixing heights limited by 50 to 100 meters, the dispersion of air pollutants is considerably restricted. For many Canadian sites, a data-collection period of twelve months would be adequate to meet these requirements. Shorter studies should only be considered if there is a requirement to assess a specific problem area or emissions source at an individual airport.

An analysis of the air pollution data collected from four major federal airports, four local airport authorities and Toronto Island (see under Carbon monoxide (CO) below) indicates that at four sites (Calgary, Dorval, Vancouver and Pearson) air quality objectives for the monitored pollutants were regularly approached and occasionally exceeded.

Ozone has only been monitored at airports since 1987. The applicable federal objectives for ozone have been regularly exceeded at all five sites that were tested (i.e. Dorval, Edmonton, Halifax, Pearson and Vancouver).

Transport Canada records show Calgary, Dorval, Vancouver and Pearson to be the busiest in terms of aircraft itinerant movements; recorded for 1992 at approximately 200, 197, 289 and 327 thousand movements, respectively.[35] This represents an increase in air traffic of approximately 3% for Calgary since the 1989 study and a decrease in air traffic of approximately 11% for Vancouver, 29.9% for Dorval and 6% for Toronto since the 1989 study. Although the relation between air pollutant levels and aircraft traffic is certainly not expected to be linear, significant increases in pollutant concentrations are probable at those sites in the future, due to intensified aircraft service vehicle and ground access traffic activities.

Due to the past recession, airline restructuring and the rationalization of the air travel industry, aircraft movements were generally lower than in previous years. With this reduction in air traffic came an improvement in ambient air quality. However, it is expected that air traffic volumes will eventually return to average predicted levels. Therefore the conclusions stated in the 1991 air quality report are still valid. At current traffic growth rates at these major airports, a monitoring program should be scheduled about every five years. The need for future air quality monitoring studies should be determined by itinerant aircraft volumes, expansion plans and new facility projects at other major Canadian airports where monitoring has occurred.

AIR MONITORING EQUIPMENT

In 1978/79, Transport Canada designed and constructed a mobile air monitoring laboratory in order to determine the air quality at major federal airports in comparison with the National Air Quality Objectives. Its purpose was to help establish the environmental impact of airport operations in Canada. In 1993, a new mobile air monitoring truck was constructed and is presently in the process of taking samples at Ottawa's Macdonald-Cartier International Airport (1994). Principle equipment being used in the air monitoring truck is listed below:

☐ carbon monoxide NDIR spectrometer
☐ nitrogen oxides duel-channel chemiluminescent spectrometer
☐ total suspended particulate hi-volume sampler
☐ PM_{10} particulate sampler
☐ flame ionization detection (FID) total hydrocarbon analyzer
☐ ozone UV absorption spectrometer
☐ meteorological monitor (air temperature, wind speed, wind direction)
☐ dynamic calibrator
☐ 20 channel multipen hybrid digital chart recorder/data logger system
☐ catalytic oxidizer
☐ zero air supply system
☐ calibration gas supply system.

DOAS

Differential optical absorption spectrometer (DOAS) has been used by a number of investigators over the past ten years to measure a wide range of gaseous air pollutants. The Differential Optical Absorption Spectrometer (DOAS) instrument has a number of features which make it attractive for field monitoring studies in remote and urban areas:

☐ it is able to measure all trace gases at the same time
☐ the detection limits are very low
☐ several individual or specific hydrocarbons can be recorded
☐ the measurement along a path seems to be more representative than a point measurement
☐ low maintenance

☐ it is a complete unit including software
☐ the system is extremely sensitive and highly accurate
☐ the results are directly displayed, as the analyzer does all the necessary calculations
☐ the DOAS system includes an easy to use statistics software package for presenting data in graphs and tables
☐ complete on-line system gives results twenty four hours a day.

DOAS system and operation

The following is a brief explanation of how the DOAS system operates. A beam of light is projected from a transmitter to a receiver over a pre-determined path which may be separated by 0.1–2000 m. As the light travels through the air, gas particles absorb particular wavelengths different gases absorb different amounts of certain wavelengths – and the receiver and analyzer/computer measure the changes in the wavelengths. The light is then sent over a fiber-optic cable to a central unit containing a computerized spectrometer, where it is analyzed to determine the type and quantity of gaseous substances present in the measurement path. The results of the measurements are directly displayed on a monitor in numerical, tabular or graphic form. DOAS can measure twelve or more different gases/pollutants (SO_2, NO_x, NH_3, Hg, CO_2, C_6H_5OH, HNO_2, Formaldehyde, Hydrocarbons, Phenol, HCl and O_3). Although the DOAS system is unable to detect carbon monoxide (CO), a carbon monoxide analyzer can be logged into the DOAS system allowing for complete on-line monitoring. Since measurements are based on variations in wavelength levels and not absolute absorption, the readings of concentrations are not affected by particles in the air.

The DOAS Air Monitoring approach has received widespread acceptance in Europe with many industrial, municipal and government installations. Currently there are over 300 systems in operation throughout Europe and the US including four airports: Zürich, Dusseldorf, Manchester and Geneva. Canada has three DOAS systems in operation at present. The effectiveness of this system is being tested at Lester B. Pearson International Airport. The results of the tests taken at LBPIA will determine whether other Canadian airports adopt the system in future.

ANALYSIS OF 1978 TO 1994 AIRPORT MONITORING DATA

Following the design, development, construction and testing of the Transport Canada Mobile Air Monitoring Laboratory in 1977/78, the system commenced a series of long-term monitoring projects at the major federal airports. Studies were conducted at the following airports:

☐ Macdonald-Cartier (July 1978 – Feb. 1979)
☐ Lester B. Pearson (Feb. 1979 – July 1979, Oct. 1983 – Feb. 1984, May 1989 – Jan. 1990)
☐ Vancouver (July 1979 – Nov. 1979, Aug. 1984 – Feb. 1985, April 1990 – March 1991)
☐ Montreal (Dorval) (Jan. 1981 – Mar. 1981, Jan. 1992 – Dec. 1992)
☐ Montreal (Mirabel) (May 1981 – July 1981)
☐ Winnipeg (July 1981 – Dec. 1981)
☐ Toronto Island (Nov. 1982 – Feb. 1983)
☐ Calgary (Aug. 1985 – Jan. 1986)
☐ Edmonton (Aug. 1987 – Jan. 1988)
☐ Halifax (Aug. 1988 Jan. – 1989).

The data from each monitoring study has been analyzed and summarized in detail in fifteen project reports prepared for Transport Canada under contract.[4] Data for this summary analysis and comparison has been compiled from these reports and is presented in Tables 1 to 10 for carbon monoxide (CO), nitrogen dioxide (NO_2), nitric oxide (NO), total suspended particulate matter (TSP) and ozone (O_3). As stated earlier, ozone monitoring has only been implemented since 1987 at Dorval, Edmonton, Halifax, Pearson and Vancouver.

For each major air pollutant, various averaging periods (i.e. maximum, max. average, average, 90 and 99 percentile concentrations) were calculated. In all cases, the recorded minimum pollutant concentrations were at or below the monitoring instruments' detection limits on certain days.

Carbon monoxide (CO)

Tables 1 and 2 summarize the recorded carbon monoxide data from the Canadian airport monitoring projects. The 1-hour Federal Air Quality Desirable and Acceptable Objectives for carbon monoxide (CO) are 15 mg/m^3 and 35 mg/m^3 respectively. Maximum 1-hour recorded carbon monoxide concentrations in the range of 3–17 mg/m^3 were observed during the 15 airport studies. The federal desirable objective was exceeded only at Pearson (1983/84), however, maximum levels at Calgary, Dorval, Edmonton and Vancouver airports approached this concentration. The 1-hour maximum acceptable air quality objective was not exceeded during any of the monitoring projects.

Table 1. Carbon monoxide (mg/m³)

Averaging period		1-hour				8-hours	
Fed. Obj. (Acceptable)		35				15	
Fed. Obj. (Desirable)		15				6	
MFA Airports	Dates	STND Percentile				Max	Avg
		Max	90%	99%	Avg Max		
Halifax	Aug. 88 Jan. 89	5.40	1.00	2.10	–	2.60	0.40
Ottawa	Jul. 78 Feb. 79	4.90	–	1.80	–	–	–
Pearson	Oct. 83 Feb. 84	17.20	3.00	5.00	–	–	–
Pearson	May. 89 Jan. 90	9.90	1.70	3.60	–	4.10	0.70
Toronto Isl.	Nov. 82 Feb. 83	4.60	1.00	2.30	1.50	2.80	0.80
Winnipeg	Jul. 81 Dec. 81	6.20	0.80	2.20	1.30	2.40	0.60

Table 2. Carbon monoxide (mg/m³)

LAA Airports	Dates	STND Percentile				Max	Avg
		Max	90%	99%	Avg Max		
Dorval	Jan. 81 Mar. 81	12.30	2.30	7.20	3.00	9.30	1.50
Dorval	Jan-92 Dec. 92	8.70	1.10	2.90	1.22	3.40	0.40
Mirabel	May. 81 Jul. 81	2.80	1.00	1.00	0.50	1.10	0.30
Calgary	Aug. 85 Jan. 86	12.00	1.90	6.10	3.20	7.90	1.70
Edmonton	Aug. 87 Jan. 88	12.10	0.70	2.10	1.10	3.00	0.50
Vancouver	Jul. 79 Nov. 79	14.30	2.80	8.00	3.20	8.50	1.60
Vancouver	Aug. 84 Feb. 85	13.90	2.80	6.80	2.40	11.00	3.00
Vancouver	Mar. 90 Mar. 91	7.80	1.80	3.90	2.00	3.60	0.70

Unavailable or incomplete data indicated as '–'

The 8-hour Federal Air Quality Desirable and Acceptable Objectives for carbon monoxide are 6 mg/m^3 and 15 mg/m^3 respectively. The maximum 8-hour carbon monoxide concentrations exceeded the federal desirable objective on several occasions at Calgary, Dorval and Vancouver International Airports. Based on the raw data of the 1983/84 Pearson study, neither the desirable objective of 6 mg/m^3 nor the acceptable objective of 15 mg/m^3 was exceeded at the airport.

Most of the high carbon monoxide concentrations measured during the airport monitoring studies were measured in the vicinity of the main terminals and terminal aprons. These concentrations were attributable, in part, to the emissions of taxiing and idling aircraft along with all the associated service vehicle and passenger access vehicle traffic in the terminal areas. As both the travelling public and airport personnel may breathe the ambient air at these locations for extended periods, further monitoring and analysis of carbon monoxide at these sites is warranted.

Nitrogen dioxide (NO$_2$)

Tables 3 and 4 summarize the recorded nitrogen dioxide data from the Canadian airport monitoring projects. The NO$_2$ Federal Ambient Acceptable Air Quality Objectives are 400 μg/m^3 for 1 hour and 200 μg/m^3 for 24 hours. These objectives were exceeded only at Dorval airport although high maximum levels were recorded at Vancouver, Pearson and Calgary airports. No desirable nitrogen dioxide objectives have been promulgated at this time except for the long-term annual average level.

Although direct NO emissions may be decreasing with newer aircraft engine designs, it is expected that these decreases will not be sufficient to offset the increased consumption of jet fuel by a larger fleet.[12] Considerable quantities of nitric oxides (NO) are expected to be produced by aircraft activities, and as this compound is a direct precursor of atmospheric nitrogen dioxide, monitoring of nitric oxide data is an important part of airport air quality evaluation.

Nitric oxide (NO)

Tables 5 and 6 summarize the recorded nitric oxide (NO) data from Canadian airport monitoring projects. Note, no environmental or health related objectives exist at present for nitric oxide. The major airport sources of nitric oxide are a combination of aircraft, service vehicle and access traffic emissions. Motor vehicle traffic emissions

Table 3. Nitrogen dioxide ($\mu g/m^3$)

Averaging period		1-hour				24-hours	Annual
Fed. Obj. (Acceptable)		400				200	100
Fed. Obj. (Desirable)		–				–	60
MFA Airports	Dates	STND Percentile				Max	Avg
		Max	90%	99%	Avg Max		
Halifax	Aug. 88 Jan. 89	116	39	64	–	41	12
Ottawa	Jul. 78 Feb. 79	115	–	68	–	–	17
Pearson	Feb. 79 Jul. 79	230	–	–	108	123	59
Pearson	Oct. 83 Feb. 84	226	98	136	–	112	54
Pearson	May. 89 Jan. 90	204	102	160	–	136	53
Toronto Isl.	Nov. 82 Feb. 83	146	32	45	59	74	33
Winnipeg	Jul. 81 Dec. 81	89	20	35	47	42	15

Table 4. Nitrogen dioxide ($\mu g/m^3$)

LAA Airports	Dates	STND Percentile				Max	Avg
		Max	90%	99%	Avg Max		
Dorval	Jan. 81 Mar. 81	411	101	200	97	210	56
Dorval	Jan. 92 Dec. 92	178	79	126	78	105	37
Mirabel	May. 81 Jul. 81	137	19	41	44	35	14
Calgary	Aug. 85 Jan. 86	250	40	61	61	73	32
Edmonton	Aug. 87 Jan. 88	137	47	70	50	47	56
Vancouver	Jul. 79 Nov. 79	216	68	113	70	100	35
Vancouver	Aug. 84 Feb. 85	333	45	86	83	192	50
Vancouver	Mar. 90 Mar. 91	146	68	98	65	77	35

Unavailable or incomplete data indicated as '–'

Table 5. Nitric oxide (μg/m³)

Averaging period		1-hour				24-hours	
Fed. Obj. (Acceptable)		–				–	
Fed. Obj. (Desirable)		–				–	
MFA Airports	Dates	STND Percentile				Max	Avg
		Max	90%	99%	Avg Max		
Halifax	Aug. 88 Jan. 89	142	16	58	–	–	6
Ottawa	Jul. 78 Feb. 79	138	–	66	–	–	5
Pearson	Feb. 79 Jul. 79	522	–	–	118	99	31
Pearson	Oct. 83 Feb. 84	811	143	381	–	196	54
Pearson	May. 89 Jan. 90	691	97	293	–	–	34
Toronto Isl.	Nov. 82 Feb. 83	319	42	160	78	105	21
Winnipeg	Jul. 81 Dec. 81	319	17	94	44	49	10

Table 6. Nitric oxide (μg/m³)

LAA Airports	Dates	STND Percentile				Max	Avg
		Max	90%	99%	Avg Max		
Dorval	Jan. 81 Mar. 81	770	129	506	153	304	47
Dorval	Jan. 92 Dec. 92	868	71	280	109	–	25
Mirabel	May. 81 Jul. 81	140	5	47	24	18	9
Calgary	Aug. 85 Jan. 86	785	82	299	158	214	35
Edmonton	Aug. 87 Jan. 88	387	120	120	57	–	9
Vancouver	Jul. 79 Nov. 79	>600	162	536	176	270	23
Vancouver	Aug. 84 Feb. 85	786	229	474	234	467	89
Vancouver	Mar. 90 Mar. 91	505	96	300	115	–	33

Unavailable or incomplete data indicated as '–'

appear to have a greater effect on landside concentrations while aircraft and service vehicle emissions effect the airside concentrations.[10] In the presence of sufficient ozone, the reaction of NO to NO_2 is relatively rapid and stoichiometric; that is, one molecule of NO_2 is produced for each precursor molecule of NO. The reaction is accelerated by warm temperatures. Thus, high nitric oxide emissions from airport operations will effect nitrogen dioxide concentrations within the airport environment and beyond. Levels of oxides of nitrogen measured at airports were comparable to, or lower than, those measured in surrounding urban areas.

Due to increasing air traffic, the utilization of high-power, high-temperature jet engines and the associated service and access vehicle activity, nitrogen oxide emissions from airports can be expected to increase at a steady rate. It should be noted that NO_x emissions are not as easy to control as those of carbon dioxide and hydrocarbons (HC), thus requiring different control strategies for NO_x than for other pollutants i.e. carbon monoxide. Although elevated NO_x levels at airports are due to a combination of airport and urban activities, continued compliance monitoring at the major airports Dorval, Pearson, Calgary and Vancouver is recommended as current nitrogen dioxide levels approach, and occasionally surpass, the Canadian guidelines.

Total suspended particulate (TSP)

Tables 7 and 8 show the maximum, standard-percentile and average total suspended particulate (TSP) concentrations from the airport air quality studies. The 24-hour Federal Acceptable Objective concentration for TSP is 120 μg/m^3. The measurement technique involves a 24-hour sampling period and is normally conducted twice per week. It can be seen that at all airports except Halifax, Ottawa and Toronto Island, TSP samples exceeded the 24-hour Federal Acceptable Objective. The percentage of samples exceeding the objective varied from 2% at Calgary and Winnipeg to 18% during one study at Pearson.

Suspended particulate concentration objectives are regularly exceeded in the industrial regions of Canada, although there has been a general decrease in TSP levels in the last ten years. Consensus was not reached in the airport study reports concerning the primary source of the higher TSP concentrations. No firm correlation could be established between TSP levels and wind direction, wind speed or airport activities. This was due to the limited number of samples, the 24-hour averaging period for each sample and the ubiquitous nature of suspended particulate sources. Examination of the sample filters

Table 7. Total suspended particulate ($\mu g/m^3$)

Averaging period		24-hours			Annual
Fed. Obj (Acceptable)		120			70
Fed. Obj (Desirable)		–			60
MFA Airports	Dates	STND Percentile Max	99%	Percent Exceeding 24-hr Obj	Avg
Halifax	Aug. 88 Jan. 89	66	49	0	16
Ottawa	Jul. 78 Feb. 79	82	73	0	36
Pearson	Feb. 79 Jul. 89	171	155	18	81
Pearson	Oct. 83 Feb. 84	147	99	3	40
Pearson	May. 89 Jan. 90	234	205	10	53
Toronto Isl.	Nov. 82 Feb. 83	71	64	0	38
Winnipeg	Jul. 81 Dec. 81	173	96	2	34

Table 8. Total suspended particulate ($\mu g/m^3$)

LAA Airports	Dates	STND Percentile Max	99%	Exceeding 24-hr Obj	Avg
Dorval	Jan. 81 Mar. 81	214	152	14	67
Dorval	Jan. 92 Dec. 92	168	150	6	32
Mirabel	May. 81 Jul. 81	132	58	5	41
Calgary	Aug. 85 Jan. 86	152	74	2	29
Edmonton	Aug. 87 Jan. 88	144	142	3	30
Vancouver	Jul. 79 Nov. 79	194	156	9	67
Vancouver	Aug. 84 Feb. 85	166	138	7	39
Vancouver	Mar. 90 Mar. 91	82	78	0	26

Unavailable or incomplete data indicated as '–'

indicated that the collected particulates originated from dust and soil as well as from smoke and other combustion products. Certain samples obtained at Winnipeg and Edmonton clearly showed soil from wind erosion whereas samples taken at Vancouver and Dorval were black and carbon-rich. Suspended matter at Pearson (1989/90) consisted of a mixture of combustion related particles and soil particles from wind erosion/construction.[14]

Ozone (O_3)

Tables 9 and 10 summarize the recorded ozone (O_3) data from Canadian airport monitoring projects. Ozone monitoring capability was acquired in 1987 and subsequently used for Dorval, Edmonton, Halifax, Pearson and Vancouver monitoring programs. The 1-hour Federal Acceptable Objective for ozone is 160 $\mu g/m^3$. Recorded ozone levels were relatively high for all five studies. Objectives were exceeded at Dorval (1992), Vancouver (1990/91) and in particular Pearson (1989/90) where objectives were exceeded on 22 of the monitoring days. Maximum values exceeding the Desirable Objective of 100 $\mu g/m^3$ were recorded at both Halifax and Edmonton.

The 24-hour Federal Acceptable Objective for ozone is 50 $\mu g/m^3$. Concentrations exceeded the corresponding Acceptable Objective level of 50 $\mu g/m^3$ at all five sites.

High ozone levels frequently occurred when other pollutant concentrations, aircraft and airport activity, were relatively low. Ozone is highly reactive and normally combines quickly with nitric oxide to form nitrogen dioxide, and also reacts with hydrocarbons in complex photochemical processes. Thus the net effect of many combustion-generated pollutants in the vicinity of airports would often reduce the local ambient ozone levels. The measured dependence of wind direction and wind speed on ozone concentrations confirms the above conclusion.[14]

STUDY OF POSSIBLE REVISIONS TO AIR QUALITY OBJECTIVES

In order to establish the necessary directions and areas of emphasis required for future airport air quality projects, a study of proposed and possible revisions to the National Air Quality Objectives was undertaken. The 1991 ambient air quality report reviewed the World Health Organization, US and Canadian standards for air quality. Also

Table 9. Ozone (μg/m³)

Averaging period		1-hour				24-hours	
Fed. Obj. (Acceptable)		160				50	
Fed. Obj. (Desirable)		100				30	
		STND Percentile					
MFA Airports	Dates	Max	90%	99%	Avg Max	Max	Avg
Halifax	Aug. 88	137	67	98	–	72	–
	Jan. 89						
Ottawa	Jul. 78						
	Feb. 79						
Pearson	Jan. 90	271	90	167	–	143	33
Toronto Isl.	Nov. 82						
	Feb. 83						
Winnipeg	Jul. 81						
	Dec. 81						

Table 10. Ozone (μg/m³)

		STND Percentile					
LAA Airports	Dates	Max	90%	99%	Avg Max	Max	Avg
Dorval	Jan. 92	184	73	129	65	97	35
	Dec. 92						
Mirabel	May. 81						
	Jul. 81						
Calgary	Aug. 85						
	Jan. 86						
Edmonton	Aug. 87	152	67	98	–	72	–
	Jan. 88						
Vancouver	Mar. 90	166	57	80	51	55	23
	Mar. 91						

Unavailable or incomplete data indicated as '–'

obtained, and reviewed, were a number of Governmental publications relating to the planning and establishment of air quality standards as well as summary tables showing the current ambient air objectives for Canada, the USA and WHO.

A comparison of these air quality criteria is shown in Table 11, for the pollutants monitored during the airport assessment projects. It can be

Table 11. Comparison of Canadian Air Quality Objectives, USA Air Quality Standards and World Health Organization Guidelines

Air Contaminant	Canada			USA		WHO Guidelines
	Desirable	Acceptable	Tolerable	Primary	Secondary	
Carbon monoxide (mg/m³)						
Average concentration over a 1-hour period	15	35	–	40	–	30
Average concentration over an 8-hour period	6	15	20	10	–	10
Nitrogen dioxide (µg/m³)						
Average concentration over a 1-hour period	–	400	1000	–	–	400
Average concentration over a 24-hour period	–	200	300	–	–	150
Annual arithmetic mean	60	100	–	100	100	–
Oxidants (Ozone) (µg/m³)						
Average concentration over a 1-hour period	100	160	300	235	235	150–200 100–120 (8 hours)
Average concentration over a 24-hour period	30	50	–	–	–	–
Annual arithmetic mean	–	30	–	–	–	–
Suspended particulates (µg/m³)						
Average concentration over a 24-hour period	–	120	400	(Note 1) 260 (150)	150 (150)	120
Annual geometric mean	60	70	–	75 (50)	60 (50)	–

Note (1) Parentheses show the USEPA AQS for PM_{10} which replaces the TSP standard

83

seen that the Canadian Maximum Acceptable Air Quality Levels, which are 'intended to provide adequate protection against adverse effects on soil, water, vegetation, materials, animals, visibility, personal comfort and well being',[36] have been set at concentrations comparable to those established by the United States Environmental Protection Agency (USEPA) and the World Health Organization (WHO).

Carbon monoxide

The 1-hour carbon monoxide objectives for Canada, USA and the World Health Organization are 35 mg/m^3, 40 mg/m^3 and 30 mg/m^3 respectively. Although Canada's objective is lower than US standards for CO, it is still marginally higher than WHO standards. If Canada was to adopt the WHO standard for carbon monoxide the consequences would be insignificant since no Canadian airports recorded levels above 18 mg/m^3.

The 8-hour Acceptable Canadian carbon monoxide objective (15 mg/m^3) is 50% higher than both the USA Primary and Secondary standard and the WHO guideline for this pollutant; this difference is significant in terms of the maximum recorded level during air quality studies at the major airports. A recent review of the National Air Quality Objectives for carbon monoxide has resulted in the recommendation that the Canadian Acceptable Level for CO for an 8-hour averaging period be reduced from 15 mg/m^3 to 12.7 mg/m^3 which is still higher than both the USA and WHO standards. The impact that this recommended change will have on airport quality assessment is minimal; no 8-hour CO measurements during the past airport studies have exceeded this concentration although it has been approached at some sites. If, however, the carbon monoxide objective is revised to be in line with USA and European guidelines (10 mg/m^3), then some exceedences are to be expected at major airports, especially in the vicinity of the main terminal aprons.

Nitrogen oxides

The Canadian Air Quality Objectives for nitrogen dioxide (NO_2) have remained since their original establishment during the 1970s. A recent consultation with the Federal-Provincial Committee responsible for updating air quality criteria has indicated that no revision is planned in the near future. While the USA completed a review of NO_2 standards in 1985, another review is currently underway with health effect studies. The US standard for the oxides of nitrogen considers only an

annual arithmetic mean for NO_2 concentrations. However, the introduction of short-term standards as well as the inclusion of standards
for nitric oxide (because of its implication in acid rain) is possible in
the future. Canada and WHO guidelines for one-hour nitrogen dioxide
concentrations are the same (400 $\mu g/m^3$) and are not expected to
change in the near future. However, the 24-hour nitrogen dioxide
WHO guideline of 150 $\mu g/m^3$ is 25% lower than the Canadian acceptable objective of 200 $\mu g/m^3$. No corresponding USA levels for NO_2
have been set but are being considered at this time; the Council of
European Communities have established a 98% limit value of 200 $\mu g/$
m^3 and a 98% guide value of 135 $\mu g/m^3$. Adoption of the WHO standard for 24-hour concentrations of NO_2 would result in exceedences at
Dorval and Vancouver International airports. Airport air quality compliance would thus be impacted somewhat if NO_2 objectives are lowered. Objectives might also be affected if nitric oxide limits are set,
since this compound is present in relatively high ambient concentrations (100 to 1000 $\mu g/m^3$) as a primary combustion product from aircraft and vehicular operations.

Total suspended particulates

Total suspended particulate (TSP) concentrations have been measured
for over 20 years by a high-volume sampling technique that was non-
selective regarding particulate size. In practice, particulate of up to
45 μm in diameter were collected and retained by the apparatus intake
and filter system. Recent evidence gathered by the USEPA indicated
that a more useful and direct health-related determination of particulate pollution would be obtained if maximum particulate sizes were
restricted to 10 μm or less. In 1987, new primary ambient particulate
matter (PM) standards were established in the USA at 150 $\mu g/m^3$ (24-
hour) and 50 $\mu g/m^3$ (annual geometric mean) to replace the previous
primary standards of 260 and 75 $\mu g/m^3$ for TSP, respectively. While
new US standards for total suspended particulates have been substantially lowered, the new objective of 150 $\mu g/m^3$ (24-hour) is still
20% higher than both the Canadian and WHO standards (120 $\mu g/m^3$).

As no data is available that would permit the estimation of particulate
matter (PM) levels at airports from previously measured TSP concentrations, suspended particulate matter compliance at airports with
relation to these new (US) standards is not predictable. However,
measurement data at some US airports (e.g. Greater Pittsburgh International) will soon be available and should provide some guidance on
the PM standard at airports. Revision of the Canadian particulate
matter objectives is likely to take some time because of the major

implications in monitoring instrument investment and data analysis/ interpretation; the present TSP objectives will probably remain in effect through the 1990s.

Ozone

Canadian standards for ozone (O_3) concentrations are comparable to WHO standards but are far more stringent than the objectives set by the USA Canadian and WHO standards for 1-hour ozone concentrations (160 $\mu g/m^3$) are substantially lower than the US objectives of 235 $\mu g/m^3$. Information received from USEPA officials suggested that ozone standards will not be relaxed. Current evidence concerning the health and environmental effects of this reactive compound indicates a need to reduce ambient concentrations of ozone.

While the results from Dorval (1992), Edmonton (1987/88), Halifax (1988/89) and Pearson (1989/90) showed ozone concentrations approaching or exceeding Federal Acceptable Levels, it was concluded from this study that the airports did not appear to be the local source of this pollutant.

Other proposed revisions

Another change to airport objectives could be the use of alternative fuels as an abatement strategy for stationary sources, service vehicles, motor vehicle fleets (i.e. buses, taxis, limos), and fire training. Inclusion of this strategy will require the use of new emission factors and different monitoring methodologies but could reduce concentrations at the critical terminal area for both the landside and airside. Monitoring of hydrocarbons could be a prime concern in the future, as it has become in Europe, due primarily to aviation fuel complaints. The monitoring and modelling of toxic emissions could also be an important future consideration.

EVALUATION OF AIRPORT AIR QUALITY MODELLING

Interest in the development of Airport Air Quality Simulation Models began in the early 1970s with the substantial increases in commercial air carrier traffic and new and expanded airport facilities. Concern over the potential environmental impact of such increased aircraft and airport-related activities prompted considerable research efforts directed at the predictive modelling of airport pollutant emissions and their subsequent dispersion in the surrounding areas. Many US airports rely simply on an emission inventory process for planning purposes and do not perform dispersion modelling until required. This is because modelling at airports is particularly complex as most of the emission sources are mobile and their activity and emissions characteristics vary considerably with diurnal, weekly, seasonal and weather-related factors. In addition, for aircraft, calculations of a three-dimensional spatial and temporal distribution of emissions are required.

The advent of rapid, powerful microcomputers spawned the development of a new series of airport air quality modelling systems, financed primarily by the Federal Aviation Administration and the US Air Force. These range from Simplex 'A', which is a special-case program designed to model the exhaust-plume emission and dispersion associated with an accelerating aircraft to the most recent Emissions and Dispersion Modelling System (EDMS) which is a more comprehensive airport air quality simulation program. The main goals addressed in the development of these programs were:

□ to permit operation on accessible, conventional microcomputers
□ to modularize the program design with separate emissions, data input and dispersion calculation sections for each type of source
□ to allow for straightforward, interactive and graphical entry of pollutant sources, source and receptor locations, emission factors
□ to combine the requirements for civil and military airport modelling, reducing redundancy and allowing application at mixed-use airports
□ to incorporate new and improved emission factors, activity descriptions and dispersion modelling equations as the knowledge base increases and
□ to permit the use of the assessment programs by non-technical personnel.

The 1991 air quality report reviewed the three recent simulation programs, Simplex 'A', the Graphical Input Microcomputer Model (GIMM) and the Emissions and Dispersion Modelling System (EDMS). The study concluded that EDMS was the most practical and applicable program for airport purposes. It must be emphasized that even with the most careful preparation of an input data base and correct execution of all operational procedures, air quality model predictions of short term and short distance pollutant concentrations often differ by factors of 2 to 10. Thus they are primarily useful as an environmental screening tool and are not intended or capable of replacing on-site compliance monitoring.

It is important to recognize that there is a great difference between 'operating' one of these complex dispersion simulation programs and actually setting up and applying a program to a real airport situation in an appropriate, scientific manner. In the preparation of an air quality impact assessment a number of judgment-calls and decisions are required that should only be made by experienced technical personnel. As a minimum, the following information is required to run the model.

☐ A detailed, scaled site-configuration; location of main emission sources (runways, taxiways, aprons, access roads, parking lots, major point sources), location of receptors or sensitive environmental areas.

☐ Operational characteristics for all sources: air and ground traffic, power and heating plants and other major point source usage. Typical air traffic operational modes; usage of runways, queuing, takeoff and landing profiles. Details for landing/take-off cycle; estimation of taxiing and idling times.

☐ Emission factors for all types of sources expected at or near the site: (aircraft by model and engine), passenger vehicles, public transit, heating plants, other stationary sources. The selection of emission factors involves knowledge of traffic mix and operating speeds combined with pollutant emission rates.

☐ Meteorological scenarios: as this is the factor affecting the dispersion of emissions, careful consideration to the choice of weather conditions has to be given. A thorough review of the airport site weather records should also be made in order to estimate variability of wind directions, wind speeds, temperatures, stability classes and mixing heights. In addition, as the selection of airport runways and taxiways is directly influenced by weather, the emissions modelling procedure must compensate for the restriction of operational scenarios for certain meteorological conditions.

☐ Current background air pollutant concentration data: even the best airport simulation programs cannot predict air quality levels with any reliability if an 'offset' for air pollution levels present in the ambient air is not included. This offset is, of course, dependent upon the same temporal and meteorologic factors that affect airport pollutant dispersion and must be appropriately corrected.

Also highly desirable in the application of an air quality predictive model at a given airport site is the gathering of actual monitoring data concurrently with the modelling work. This permits a comparison of simulations using the program for actual, air and vehicle traffic, meteorologic conditions and background pollution levels with measured concentrations. Unfortunately, a recent survey of US airports showed the modelling efforts are approximately twice that of the monitoring efforts.

As the development of the three microcomputer models has been evolutionary, culminating in the present EDMS program, most features available in the earlier Simplex 'A' and GIMM programs have been upgraded and incorporated into EDMS. Due to significantly more powerful hardware and memory capability of the computers used by EDMS, this program was the only one undergoing continuing development. In the past, EDMS had been used for the preparation of environmental impact statements (EIS) at Vancouver International and Pearson International. Presently, EDMS is primarily used by airports as an environmental screening tool to prove that runway expansions will not increase air pollution. EDMS has recently received approval from the United States Environmental Protection Agency as a preferred screening model.

CONCLUSIONS AND RECOMMENDATIONS

It is clear from the evidence in the preceding sections that air quality at major international airports, while not generally worse than that found in the surrounding metropolitan areas, is measurably impacted by the activities of aircraft, ground and service vehicle traffic and other major emission sources located on the airport grounds etc.

Transport Canada has completed monitoring projects at all international airports, thus acquiring quality baseline data that can be used in master planning activities, environmental impact studies and as an aid in evaluating long-term quality trends at these airports. It is evident from the monitoring data that the National Air Quality Objectives for the gaseous pollutants carbon monoxide and nitrogen dioxide are in danger of being regularly approached or exceeded only at the most active airports (Calgary, Dorval, Pearson and Vancouver) within the next few years. TSP objectives are occasionally exceeded at nearly all airport sites although the sources of these high values have not been traced to particular aircraft or airport emissions. Ozone levels have only been monitored during the most recent studies at Dorval, Edmonton, Halifax, Pearson and Vancouver where short term concentrations frequently exceeded the maximum desirable air quality objectives.

In order to ensure that the National Air Quality Objectives are being met where possible at the major federal airports, as well as to access air pollution trends and potential mitigational measures at these sites, it is recommended that Transport Canada undertake the following.

1. Retain the Department's current capability to undertake air monitoring by maintaining the existing Mobile Air Monitoring System (MAMS).

2. Continue the practice of carrying out approximately one planned airport monitoring study per year; in order to access air quality adequately it is necessary to monitor over as many different seasons as possible: twelve months would be the ideal duration.

3. Schedule monitoring at the most active airports, where air quality has already been determined to approach or exceed federal objectives, at five-year intervals. This would allow the establishment of trends with reference to data gathered in the initial (baseline) and following studies.

4. Continue to concentrate monitoring activity on a terminal apron and other central areas including the landside of the terminal. Although the monitoring studies undertaken to date show that communities adjacent to the airports do not generally have their air quality adversely affected by airport generated pollutants it is clear that at certain on-airport locations, high concentrations, particularly of CO, NO and NO_2, exist due to aircraft, service vehicle, access traffic and other sources near the main terminal areas. Since both the travelling public and airport workers may be exposed to these elevated levels, it would be useful to increase the amount of monitoring data collected at these locations. The terminal area is also one of the few parts of an airport where practical pollution mitigational measures may be possible employing techniques such as the use of electrical service vehicles, control of access traffic and the towing of aircraft (reducing power-backs etc.).

5. In 1990 a THC hydrocarbon sampler was acquired and implemented in the monitoring study of Vancouver International (1990/91). Alternative measurement techniques such as the DOAS system, which was recently installed at LBPIA, should be considered in helping to monitor hydrocarbons. Also to be considered is the possibility of acquiring a mobile DOAS system to complement the instruments in the mobile air monitoring truck. Given the substantial levels and diversity of hydrocarbons that are likely to be detected, a lack of specific hydrocarbon information may raise more questions. Therefore consideration should be taken in using the DOAS system to help distinguish the source, location and type of hydrocarbons being released.

6. The present Environment Ontario and Environment Canada posi-

tion on suspended particulate matter is to retain the current 24-hour maximum acceptable concentration of 120 $\mu g/m^3$ and annual geometric mean concentration limits of 60 to 70 $\mu g/m^3$ respectively. If there is interest in relating ambient TSP levels to local sources, it is recommended that particulates be monitored on a continuous basis allowing for short-term (i.e. two-minute) averages.

7. Continue to relate the results of airport air quality studies to the applicable objectives and standards and to the air quality of the surrounding metropolitan areas. While it may be established that airports were 'good neighbours' from an air pollution point of view during the last decade, continuing tightening of standards and the abatement of urban emissions from non-airport sources combined with steadily increasing emissions from air traffic and associated airport activities may upset this balance.

8. Acquire the computer simulation programs and documentation necessary for airport air quality modelling. Significant advances in the microcomputer field have made air quality modelling more practical and accessible for the general user. It was concluded from a review of six airport-specific models that the Emissions and Dispersion Modelling System (EDMS), produced under joint development of the USA Federal Aviation Administration and the US Air Force is the most practical and applicable program for these purposes currently available. Its major application is found in preparing environmental impact assessments for new airport facilities and major expansion projects. Calibration of an air quality model for a particular airport would best be accomplished by conducting a monitoring study concurrently with the computer modelling work.

9. Since Environment Canada performs all the calibration tests on the reference gases used for the instruments in the air monitoring truck they can verify that the figures recorded are exact. Therefore it is important to continue the relationship with Environment Canada as an independent outside party. This will eliminate any questions asked on the validity of the figures recorded by the air monitoring truck.

10. In order to obtain effective air monitoring results, testing periods of twelve months are needed. Due to the fact that Transport Canada has only one mobile air monitoring truck, the time period between monitoring studies at the same airport could be several years. Therefore it is suggested that the larger Canadian airports move to a fixed monitoring system such as the DOAS system being used at Lester B. Pearson International. This would allow for continuous on-line monitoring of air quality at all the major airports. However, it should be noted that the DOAS system is

used to monitor ambient air quality only. Certain areas within an airport's boundaries are not appropriate for monitoring using this system. Therefore, all airports should continue to be monitored using the mobile air monitoring truck as well.

11. In an attempt to keep monitoring equipment up to date, it is recommended that the following outdated equipment be replaced: carbon monoxide analyzer, meteorological instrument for monitoring air temperature, wind speed and wind direction, and a zero air supply system.

12. Presently, ozone level estimation in the mobile air monitoring truck is performed using an internal supply system. These recorded levels must then be confirmed with Environment Canada's recorded ozone levels. It is recommended that a portable ozone calibration source be purchased for the MAMS to estimate atmospheric ozone levels. The new system will not only increase the accuracy of the ozone levels but it can also be verified by Environment Canada who would calibrate the new system.

13. Some new concerns may also need to be addressed at airports, these include: use of alternative fuels as a mitigation methodology for service vehicles, stationary sources, motor vehicle fleets, and fire training; the measurement and modelling of toxic emissions; and new control strategies for NO_x. Indoor air quality may also be a future concern at airports as well as ozone depletion at cruising altitudes.

It is felt that the implementation of the above recommendations would further enhance the air quality studies at Transport Canada airports. These recommendations would expand the monitoring capabilities and increase the accuracy of assessment.

REFERENCES

1. *Review of Ambient Air Quality at Major Canadian Airports*, Report #TP 9609, Tradewind Scientific Limited, December, 1988.

2. *Environmental Protection Agency Proposed Revisions to Gaseous Emission Rules for Aircraft & Aircraft Engines*, Federal Register, 43FR12615 (1978).

3. *Committee on Aircraft Engine Emissions; Final Report of Working Group A*, International Civil Aviation Organization, CAEE/2 (1980).

4. Analysis of Air Monitoring Results, Individual Airport Reports prepared for the Environment Review Services Division, Transport Canada, by the Ontario Ministry of the Environment, Trade-

wind Scientific Ltd. and Environmental Applications Group Ltd. (15 Reports, 1978–90).

5. *Air Quality and Heathrow; Results of Air Monitoring*, London Scientific Services & Environmental Health Department, London Environmental Supplement No. 15 (1987).

6. *A Microcomputer Pollution Model for Civilian Airports and Air Force Bases; User's Guide*, United States FAA, Report No. FAA-EE-85-4 (1985) (Including Updates and Instruction Memoranda through Version 3 (1988).

7. Aircraft and Air Pollution, *Environmental Science and Technology*, Vol. 15, No. 4 (April 1981).

8. *Impact of Aircraft Emissions on Air Quality in the Vicinity of Airports, an Updated Model Assessment of Aircraft Generated Air Pollution at LAX, JFK and ORD*, Report FAA-EE-80-09B, Vol. 2 (July 1980).

9. Air Pollution Associated with Airports – A Case Study, *Environmental Technology Letters*, Vol. 7, No. 4 (April 1986).

10. The Influence of Aircraft Operations on Air Quality at Airports, *Journal of the Air Pollution Control Association*, Vol. 31, No. 8 (August 1981).

11. *Airport Air Quality Policy, Modeling and Monitoring, Project Reference Number 90-180-6*, Wayson, R.L., Bowlby, W., Vanderbilt University, November 1, 1990.

12. *New Estimates of NO_x From Aircraft: 1975–2025*, Kavanaugh, M., 81st Annual Meeting of APCA, Dallas, Texas, June 19-24, 1988.

13. *Halifax International Airport Ambient Air Quality Monitoring Program (August 1988 – January 26, 1989)*, Tradewind Scientific Ltd., Report #TS-175, April 1989.

14. *Analysis of Air Monitoring Results Toronto (Lester B. Pearson) International Airport (May 2, 1989 – January 29, 1990)*, Tradewind Scientific Ltd., Report #TS-178, April 1990.

15. *Stopping Air Pollution at its Source; Discussion Paper*, Clean Air Program, Environment Ontario, Regulation 308 (1987).

16. *State of the Environment Report for Canada*, Environment Canada, Document No. EN 21-54/1986E (1986).

17. *Federal Regulatory Plan 1988*, Office of Privatization and Regulatory Affairs, Communications Directorate, No. BT57-2/1988E (1987).

18. *An Action Plan for Environmental Law Enforcement in Alberta*, Review Panel on Environmental Law Enforcement (1987).

19. *Review of National Ambient Air Quality Objectives for Carbon Monoxide & Review of National Ambient Air Quality Objectives for Sulphur Dioxide*, Federal-Provincial Committee on Air Quality, Nos. En-42-17/1-1987E & En-42-17/2-1987E (1987).

20. *Clean Air Act: An Overview*, Environment and Natural Resources Policy Division, Congressional Research Service (March 1987).
21. *London Environmental Bulletins*, Greater London Council & the London Residuary Body. Vols 1–4 (1984–98).
22. A Plenary Review of The Clean Air Act is Urgent, *Public Utilities Fortnightly*, (March 1986).
23. Regulations for Implementing Revised Particulate Matter Standards: Proposed Rule, *Federal Register*, Vol. 50, No. 63 (April 1985).
24. Retention of the National Ambient Air Quality Standards for Nitrogen Dioxide; Final Rule, *Federal Register*, Vol. 50, No. 118 (June 1985).
25. Fuel Odours Data Needed, *Airports International* (December 1986/ January 1987).
26. *Impact of Aircraft Emissions on Air Quality in the Vicinity of Airports, Vol. 1: Recent Airport Measurement Programs, Data Analyses' and Sub-Model Development & Vol. 3: Air Quality and Emission Modelling Needs*, Federal Aviation Administration Reports No. FAA-EE-80-09A/C (1980 & 82).
27. *Air Quality Procedures for Civilian Airports and Air Force Bases*, Federal Aviation Administration, Report #FAA-EE-82-21 (1982).
28. *The Weather Almanac*, Fifth Edition, Gale Research Co. (1987).
29. *Air Quality Guidelines for Europe*, World Health Organization, Regional Office for Europe (1987).
30. *A Generalized Air Quality Assessment Model for Air Force Operations (AQAM)*, USAF Report #AFWL-TR-74-305 (1975).
31. *Airport Vicinity Air Pollution Model User's Guide (AVAP)*, Federal Aviation Administration Report No. FAA-RD-75-230 (1975).
32. *Airport Vicinity Air Pollution Model: Abbreviated Version User's Guide*, National Technical Information Service Report No. AD-AO61 854/65L (1978).
33. *Simplex 'A' – A Simplified Atmospheric Dispersion Model for Airport Use*, Federal Aviation Administration Report No. FAA-EE-81-8 (1981).
34. Microcomputer Graphics in Atmospheric Dispersion Modelling, *Journal of the Air Pollution Control Association*, Vol. 23, No. 6 (June 1983).
35. *Aircraft Movement Statistics*, Aviation Statistics Center, Statistics Canada Annual Report, TP 577, (1991).
36. *Canadian Environmental Protection Act*, Environment Canada, (1988) .
37. *A Review of Ambient Air Quality at Major Canadian Airports*, Report TP # 9609, Rowan Williams Davies & Irwin Inc., (1991).

AIRPORT AIR QUALITY MONITORING BY AEA TECHNOLOGY

K. J. STEVENSON, AEA Technology, National Environmental Technology Centre, Culham, Abingdon, Oxfordshire OX14 3DB, UK

This paper briefly reviews the air quality monitoring and modelling activities undertaken at UK Airports by AEA Technology. Two recent studies are used as examples

- [ ] a monitoring and modelling study at London Gatwick Airport
- [ ] the current programme of air quality monitoring at London Heathrow and London Gatwick Airports.

The London Gatwick study shows how passive diffusion tube samplers can be used to map NO_2 concentrations throughout the airport and surrounding areas. In addition, the modelling undertaken within this study has been used to provide an assessment of the percentage contribution of on-airport sources to ambient concentrations, within and outside of the airport boundary.

The scale and objectives of the current monitoring programme at London Heathrow and London Gatwick are presented. In particular, the data handling scheme used to provide real-time data to the Airport Environmental Departments and the general public is reviewed.

INTRODUCTION

This paper presents and discusses air quality monitoring and assessment studies undertaken by the National Environmental Technology Centre and the former Warren Spring Laboratory. The National Environmental Technology Centre was formed at AEA Technology, in 1994, by the amalgamation of Warren Spring Laboratory and the environmental divisions of AEA Technology.

The paper will give a general overview of these monitoring and assessment studies (Table 1); two of which have been selected for more detailed discussion.

Table 1. Airport air quality studies

1979	London Gatwick	North Terminal
1981	London Stansted	Replacement Terminal
1990	London Gatwick	Study for UK DTI
1993	Manchester	Second Runway
1989 →	London Heathrow	Terminal 5
1992 →	London Heathrow and London Gatwick	On-going monitoring

The studies at London Gatwick[1] and London Stansted[2] Airports in 1979 and 1981 served to establish a protocol and methodology for air quality assessment studies for major developments at UK airports. This protocol involves the development of current and future case emission inventories for the airport and surrounding areas, which are used for air quality modelling. The model calculations are then validated, for the present case, against on-airport and perimeter air quality monitoring.

For the Public Inquiry on the proposed second runway at Manchester Airport,[3] AEA Technology undertook a detailed modelling assessment and provided additional perimeter air quality monitoring, to supplement the extensive on-airport monitoring already being carried out by the Airport.

The 1990 study at London Gatwick[4] was not linked to a particular development, but aimed at providing a more detailed study of air quality at a major airport, with particular reference to the relative importance of the various emission sources. This study will be discussed in more detail later in this paper.

AEA Technology undertook the air quality assessment for the proposed Terminal 5 development at London Heathrow,[5] and further work continues on this study. As the Public Inquiry for Terminal 5 is currently sitting, this work will not be discussed further. The long-term air quality monitoring currently being undertaken by AEA Technology for London Heathrow and London Gatwick Airport will, however, be discussed below.

LONDON GATWICK AIRPORT STUDY 1990–91[4]

Table 2 summarises the aims of this study and the methodology utilised. The study concentrated on the investigation of the pollutant

Table 2. London Gatwick Airport Study 1990–91

Aims
☐ Assess current air quality at the airport
☐ Identify emission sources

Pollutant
☐ NO_x

Methodology
☐ Automatic monitoring	1 site
☐ Passive sampling	100 sites
☐ Gaussian modelling	

gases in the oxides of nitrogen (NO_x) family since, at the time of the study, aircraft engine emission limits for NO_x were under review.

Automatic monitoring of nitric oxide (NO) and nitrogen dioxide (NO_2) were undertaken at an on-airport location close to the end of the runway, in order to obtain detailed hourly measurements of these pollutants. In addition, passive diffusion tube samplers were used at over 100 monitoring locations to determine the spatial variation of NO_2 throughout the airport. The results from the diffusion tube survey have been interpolated to produce the contour plot of average NO_2 concentrations shown in Fig. 1. The plot shows clearly that the highest average NO_2 concentrations were centred on the Terminal area. The influence of major roads in the area is evident, as higher concentration contours are extended to the west and to the south. The survey showed little evidence of the runway itself being a significant source, although there is a slight extension of the contours at its eastern end.

The relative importance of the sources of NO_x was investigated by the compilation of a detailed emission inventory and the use of this within gaussian dispersion models to calculate ambient air pollution concentrations at points in and around the airport. As expected, a similar spatial pattern was observed for the modelled NO_x concentrations as for the measured NO_2' concentrations shown in Fig. 1.

The use of dispersion modelling allows the percentage contribution from each source to the ambient concentrations at any point to be

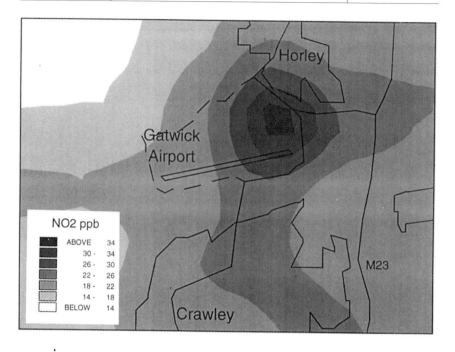

Fig. 1. Average NO$_2$ concentrations – London Gatwick Airport Study 1990/91

determined. This information is summarised in Fig. 2 for a series of points lying along a south–north line commencing close to the town of Crawley to the south of the airport (left-hand side of Fig. 2), passing through the terminal area of the airport and ending at Horley, to the north of the airport (right-hand side of Fig. 2). The lowest section of the bars representing NO$_x$ concentrations shows the contribution for the UK background; this is seen to increase for more northerly locations, as these are closer to the conurbation of London. The transect chosen passes through the area where the contribution from aircraft, shown in black on the figure, is highest, reaching about 50% of the total annual average NO$_x$ concentration close to aircraft sources. However, over most of the rest of the airport, the proportional contribution from air-craft is smaller and outside of the airport boundary, along this transect, the airport/aircraft contribution falls rapidly to a small percentage of the total (at most 10–13% just beyond the airport boundary on this north/south line). Hence, this analysis shows that aircraft are, over-all, a relatively small contributor to ambient NO$_x$ concentrations outside of the airport boundary, compared to regional and UK background sources.

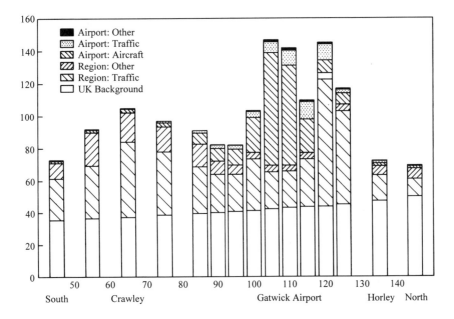

Fig. 2. Analysis of source contributions to modelled NO_x concentrations ($\mu g/m^{-3}$) along a south/north line through the main terminal area at London Gatwick Airport

MONITORING PROGRAMME AT LONDON HEATHROW AND LONDON GATWICK AIRPORTS

Table 3 outlines the aims, methodology and pollutants monitored in this long-term monitoring programme.

In order to provide real-time data, automatic analysers are sited in a mobile trailer unit at both London Heathrow and London Gatwick Airports. To ensure high quality data, documented operational procedures are used for site operation, traceable gas standards are used for calibration and all data are carefully screened prior to final archive. An outline block diagram of the data system is shown in Fig. 3. Data from the sites are collected via telemetry twice daily. After screening, they are relayed by telemetry in turn to PCs in the Environment Departments of BAA, BAA Heathrow and BAA Gatwick. This telemetry link also allows the airport staff to browse the database containing all of the data collected in this programme, which commenced in 1992.

Table 3. Monitoring programme at London Heathrow and London Gatwick

Aims
☐ Comparison with Air Quality Standards
☐ Real-time data display

Pollutants
☐ NO_x, CO, (PM_{10} and VOCs at London Heathrow), wind speed, wind direction and temperature

Methodology
☐ Automatic analysers
   – NO_x, CO, PM_{10}
☐ Passive and grab sampling
   – VOC

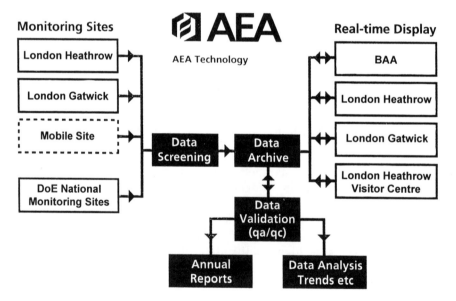

Fig. 3. Air quality monitoring at London Heathrow and London Gatwick airports – data flow diagram

In addition to the on-line access, the airports are supplied with quarterly summary reports and a more detailed annual report each year. Data from the monitoring programme are published in the regular Environment Reports[6,7] produced by both BAA Heathrow and BAA Gatwick airports.

A very brief summary of data for 1993 and 1994 is shown in Table 4. At London Gatwick airport, the annual average NO_2 concentration was 22 ppb in both years; this is approximately 75% of the NO_2 concentration in Central London. At London Heathrow, the annual average NO_2 concentration is about 30 ppb, approximately 90% of that recorded in Central London. At London Heathrow, the EC Directive Guide Value concentration was exceeded, but not the statutory EC Directive Limit Value. This guide value is widely exceeded in urban areas of the UK.

During 1995, real-time air quality data for London Heathrow have been made available to the public via an interactive on-line display at the BAA Heathrow Airport Visitor Centre. Latest concentrations of NO_2 and also carbon monoxide (CO) are shown, alongside corresponding measurements in other UK cities. An example of the screen display is shown in Fig. 4.

Table 4. Data summary NO_2

	Annual average	Comparison with standards	Comparison with other sites
London Gatwick – North Terminal 1993	22 ppb	EC/WHO not exceeded 3 hrs POOR	≈75% of London
1994	22 ppb	EC/WHO not exceeded no hrs POOR	≈75% of London
London Heathrow – Old Apron 1993	30 ppb	EC guide exceeded 4 hrs POOR	≈90% of London
1994	32 ppb	EC guide exceeded 31 hrs POOR	≈90% of London

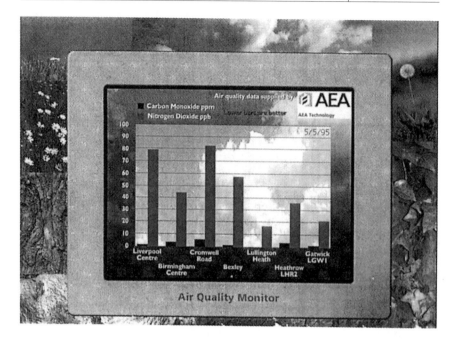

*Fig. 4. Air quality monitoring at London Heathrow and London Gatwick airports –
Heathrow Visitor Centre air quality display screen*

ACKNOWLEDGEMENTS

The 1990/91 survey at London Gatwick airport was sponsored by the
UK Department of Trade and Industry. The work at Manchester Airport was sponsored by Manchester Airport Plc. All other studies have
been fully funded by BAA plc, BAA Heathrow or BAA Gatwick Airports.

REFERENCES

1. Williams *et al. Air Pollution at Gatwick Airport – A Monitoring/
   Modelling Study in Connection with Airport Developments Proposed by BAA*. Warren Spring Laboratory 1980, Report LR 343.
   Available from AEA Technology.
2. Williams *et al. Air Pollution at Stansted Airport – A Monitoring/
   Modelling Study in Connection with Airport Developments Proposed by BAA*. Warren Spring Laboratory 1981, Report LR 386.
   Available from AEA Technology.
3. Manchester Airport Runway 2 Environmental Statement. July
   1993. Manchester Airport plc.

4. Williams M. L., Leech P. K., Stevenson K. J. and Sweeney B. *Oxides of Nitrogen Concentrations at Gatwick Airport – A Measurement and Modelling Study*. Warren Spring Laboratory 1993, Report LR 934. Available from AEA Technology.
5. Terminal 5 Heathrow Environmental Statement. February 1993. BAA plc.
6. Environmental Performance Report. BAA Heathrow Airport Ltd.
7. Environmental Progress Report. BAA Gatwick Airport Ltd.

Surface water contamination caused by airport operations

D. J. GRANTHAM, Director, Environmental Consultants & Analysts (ECAL), Hollibury House, 47 Park Street, Southport, Merseyside PR8 1HY, UK

Surface water (rainwater runoff/stormwater) is the precipitation (rain, snow, etc.) discharged from a catchment area into a stream, drain or sewer. The rate of stormwater runoff depends upon the intensity of the rainfall, and the impermeability of the catchment. The quality of the rainwater runoff reflects the nature of the activities undertaken on the catchment. Flows in stormwater sewers will also contain an element of groundwater that has seeped into the sewer.

The management of surface water discharges is regulated by increasingly rigorous legislation. Consequently the cost of statutory compliance is becoming increasingly expensive both in terms of the investment in improved pollution control infrastructure, and because of the associated constraints on operational practices. The cost of the fines for illegal discharges is also rising, although the economic consequences of adverse publicity associated with a prosecution can be a greater financial penalty.

The management of water discharges from an airport requires appropriate sewerage infrastructure, and the adoption of a comprehensive range of operational procedures. The preferred drainage strategy for a given site will be that which achieves discharge consent compliance most economically while offering the greatest protection from civil and statutory liability.

THE ENVIRONMENTAL IMPACT OF POLLUTED SURFACE WATER DISCHARGES

Vehicle and aircraft washing

The main constituents of washwater include: detergent formulations, solids, oils, greases, carbon residues, solvent residues, and heavy metals. This combination creates an effluent with a high Biochemical

Oxygen Demand (BOD), and high Suspended Solids (SS) concentrations. In addition some detergents may contain phosphates, which can create nutrient enrichment problems in receiving waters (see Appendix A).

The pollution strength of a washwater may increase as it passes through a sewerage system. This is because the detergents will emulsify oils and solubilise some solids trapped within the sewerage system.

Vehicle maintenance

Vehicle maintenance uses a wide variety of hazardous chemicals, from synthetic lubricating oils and hydraulic oils to chemicals used in plating and paint stripping operations. Aircraft maintenance makes extensive use of solvents, and can generate large quantities of waste kerosene.

The storage, handling, and disposal of the hazardous chemicals used and generated by aircraft and vehicle maintenance is heavily regulated. Consequently these materials should be absent from water discharges. Where such chemicals do contaminate the water discharges, e.g. by accidental spillage, they can have adverse environmental impacts even in low concentrations. Metal finishing wastes, and degreasing solvents in particular can be highly toxic at low concentrations in water.

Agricultural and horticultural activities

The appropriate management of agricultural land may require the use of fertilisers, creating the potential for contaminated surface water runoff. The maintenance of landscape areas may involve the use of pesticides and herbicides. The chemicals used for agricultural and horticultural purposes will generally be present in surface water discharges in very low concentrations. Larger concentrations are associated with accidental spillages.

Fuel and chemical spillage

In view of the large amounts of fuel used, a limited amount of fuel spillage is inevitable. Spillages are usually caused by human error or leakage from faulty or poorly maintained equipment. Hydrocarbon contamination of water discharges is a cause for concern because of:

☐ the high BOD of hydrocarbons
☐ the toxicity of the light oil fractions (e.g. kerosene) on the aquatic environment
☐ the fouling of the banks of the surface water discharge channels, and impounding lagoons
☐ the reduced reaeration capability of waters with oils on their surface.

In theory, any and every fuel and chemical could be spilt and eventually contaminate a surface water discharge. The environmental impact of a spillage will reflect the nature of the chemical spilt, and the measures taken to deal with the spill.

The top soil layers can become contaminated with spilt chemicals. The soil contamination can upset the soil ecosystem, inhibiting soil-borne organisms, and/or preventing the use of the soil for crop growth. In addition water percolating through contaminated soil can entrain or dissolve the contaminants, resulting in contaminated groundwater. Groundwater contamination can migrate to surface waters, creating a diffuse source of pollution in the surface water.

Emergency services

Commercial airfield licensing requirements specify the provision of adequate on-site training of all fire crews. Live fire drills will usually involve the extinguishing of simulated aircraft fires. The regular testing of crash tenders requires that fire fighting foams are dispensed on a frequent basis irrespective of live fire drills. Fire training exercises generate oil and synthetic foam residues. The effluents from fire training activities typically have extremely high BOD and high SS concentrations (see Appendix A).

Cold weather operations

Anti-icing and de-icing chemicals deployed during cold weather operations can contaminate surface water discharges. The most common de-icing and anti-icing chemicals used include:

☐ road salt (sodium chloride)
☐ urea prills (granular solid)
☐ ethylene (and diethylene) glycol
☐ monopropylene glycol
☐ calcium magnesium acetate (granular solid)

☐ potassium acetate
☐ sodium formate.

The gritty and corrosive nature of road salt limits its use to non-aircraft areas.

Glycol, acetate and formate based de-icing and anti-icing chemicals have relatively high BODs which can cause serious deoxygenation problems in receiving waters.

Urea prills dissolve to form a urea solution which exerts negligible BOD and is not toxic to fish at concentrations up to 30,000 mg/l. However microbes native to the soil and the aquatic environment readily hydrolyse urea to form ammoniacal nitrogen. Ammoniacal nitrogen has a high oxygen demand and is toxic to aquatic life. The subsequent biological oxidation of ammoniacal nitrogen produces nitrates, which can create eutrophication problems in receiving waters, resulting in secondary pollution effects. In addition both ammoniacal nitrogen and nitrate are detrimental to waters abstracted for potable supply. Thus although urea is fairly harmless, its decomposition products can have a very detrimental impact on the aquatic environment.

SURFACE WATER DISCHARGE MANAGEMENT PRACTICE

Statutory compliance with the increasingly strict surface water discharge standards requires a large capital investment in appropriate infrastructure. Of all of the environmental control programmes undertaken surface water discharge management usually requires the greatest capital investment, it typically incurs the greatest running costs, and carries some of the greatest liabilities. The expense and the liabilities associated with surface water discharge management reflects the fact that surface water pollution problems have the potential to be engineered out.

Surface water sewerage systems will discharge either directly into receiving waters or indirectly via a public stormwater sewer. Conventional drainage design allows waste materials on the catchment area, e.g. spilt fuels and de-icing chemicals, to be flushed into surface-water sewers and be discharged offsite. Consequently contaminated stormwater discharges can potentially be more grossly polluted than discharges of domestic sewage. Compliance with statutory discharge standards typically requires the control and/or treatment of surface water prior to discharge offsite.

Perennial operations

Airports constitute large impermeable catchment areas, the surface water runoff from which will often require flow balancing to mitigate potential downstream flooding problems. Flow balancing lagoons are usually used to reduce the peak surface water contamination levels (hydraulically and biologically), in addition to reducing peak flow rates offsite.

The effects of fuel and oil spillages can be abated by incorporating fuel/oil separators into the surface water sewerage systems. Oil/water separators enable separable oils to be removed from the runoff water. However conventional oil/water separators, e.g. American Petroleum Institute (API) Interceptors, and Tilted Plate Separators, will not intercept emulsified oils. Therefore it is essential that washwaters (and detergents generally) are prevented from contaminating surface water discharges.

Conventional oil/water separators are designed on the basis that oils form globules in water that eventually float to the surface from where the oil can be skimmed. Their design is based on a particular globule size (150 micron API, 60 micron Tilted Plate). Many fuels and oils used at airports form much smaller globules in water and/or associate with suspended particles, and are thus able to pass through these devices. Thus oil/water separators are merely a line of defence against fuel/oil spills and not a comprehensive treatment system.

The occurrence of fuel spills can be minimised by the following.

☐ The provision of appropriately designed centralised vehicle refuelling facilities which enable fuelling spillages to be contained within a manageable area; the drainage from which discharges to a foul sewer via an oil/water separator.

☐ Recharging the cost of a fuel spill clean up operation to the company responsible for causing the spillage. Most airports have standard operating procedures for dealing with spillages. Small spills are usually cleaned up using absorbent materials, e.g. absorbent fabrics, while larger spills are recovered, where possible, and the contaminated area cleaned up. Recharging clean up costs provides incentives to operators to minimise the incidence of fuel spills.

Surface water contamination caused by the washwater from vehicle and aircraft washing is usually prevented by providing dedicated washing areas. The specially designed wash areas drain (usually via pretreatment) to foul sewer. Thus the washwaters are dealt with as a trade effluent. The effectiveness of this approach relies upon the prevention of washing in non-designated areas.

Domestic sewage is collected by foul sewers. An airport's foul sewerage system normally discharges into public foul sewer prior to treatment at a municipal sewage treatment works. A charge is levied by the Sewerage Undertaker for accepting and treating the sewage and trade effluent. Some airports have their own sewage treatment works, the effluent from which will typically be discharged directly into a receiving water.

Cold weather operations

It is difficult and expensive to control water pollution caused by cold weather operations. The options available for reducing and ultimately abating the anti-icing and de-icing chemical contamination of stormwater runoff include:

☐ the adoption of an anti-icing strategy
☐ minimising the use of anti-icing and de-icing chemicals
☐ centralised or remote de-icing
☐ suction vehicles
☐ the use of low pollution strength anti-icing and de-icing chemicals
☐ flow balancing/first flush diversion
☐ off-site treatment
☐ on-site treatment.

Adoption of an anti-icing strategy for runways, taxiways and aprons

This helps to minimise the use of de-icing chemicals because several times the quantity of chemicals may be required to clear ice (depending upon thickness) and to maintain ice free conditions than if ice formation is prevented initially. An anti-icing strategy relies on the use of accurate ice prediction systems.

Minimising the use of anti-icing and de-icing chemicals

Operational practices that will minimise the usage of anti-icing and de-icing chemicals include:

☐ the use of equipment that will dispense optimum chemical mixes at controlled rates
☐ the correct supervision and training of staff

☐ the use of external heating for aircraft e.g. hot water de-icing, hot air de-icing, infra red de-icing or underground heating for hard-standing areas.

Centralised or remote aircraft de-icing

Centralised or remote de-icing areas can be isolated from an airport's surface water drainage system, thus preventing aircraft de-icing operations from contaminating surface water discharges. This also enables glycol based de-icing chemicals to be recovered and reused (following distillation and mixing with new feed stock). Centralised or remote aircraft de-icing is particularly suited to small or medium sized airports. Its use at large airports tends to create prohibitive operational problems.

There are two types of glycol based de-icing chemicals used on aircraft; Type 1 and Type 2. Type 1 de-icing chemicals behave like an 'ordinary' liquid. However, Type 2 de-icants are thickened to make them sticky. This enables them to maintain ice free conditions on an aircraft for longer.

Unfortunately Type 2 liquids are more difficult to recycle than Type 1. Thus centralised and remote de-icing operations select for Type 1 chemicals. This in turn selects for the de-icing facilities to be situated near to the runway, which in turn creates potential operational problems.

Stand cleaning using suction vehicles

The de-icing chemical residues left on aircraft stands following an aircraft de-icing operation can be recovered using vacuum suction vehicles. Thus preventing most of the residues from contaminating surface water.

The recovered material is sufficiently concentrated to make distillation and reuse of the (glycol based) de-icing chemicals viable.

The use of low pollution strength anti-icing and de-icing chemicals

Acetate based anti-icing and de-icing chemicals typically have a lower biochemical oxygen demand (BOD) than glycol based chemicals. Therefore the replacement of glycol based chemicals for the anti-icing and de-icing of runways, taxiways, and aprons will result in lower BOD strength stormwater contamination. Sodium formate anti-icing and de-icing chemicals have been found to be even more environmentally benign than the acetates, and have operational advantages. The adverse environmental impact of urea has caused it to be phased out at many airports.

Flow balancing/first flush diversion

An airport's surface water flow balancing lagoons can be made to reduce the peak surface water contamination levels (hydraulically and biologically) in addition to reducing peak flow rates. However this may create unacceptable odour and bird nuisance problems.

The flow balancing system can be modified so as to transfer the most contaminated flows (assumed to be the first flush) to an impounding lagoon. The grossly contaminated surface water runoff, once impounded, can then be either treated on site or delivered to an off-site treatment facility.

Off-site treatment

The biological treatment of anti-icing and or de-icing contaminated surface water runoff can be successfully achieved by admixture with sewage at a large municipal sewage treatment works. The contaminated runoff is typically collected in a first flush diversion system's impounding lagoon (see previous section). It is then discharged to a sewage treatment works in accordance with trade effluent discharge conditions. The trade effluent discharge conditions will limit the flow rate and maximum de-icant concentration of the contaminated surface water discharge. Flow rate is limited so as to not dilute and lower the temperature of the sanitary sewage, and de-icant concentrations are limited to prevent interference with the biological treatment process.

On-site treatment

☐ Most conventional physio-chemical waste water treatment methods that have been considered for de-icing contaminated runoff have been found to be unsuitable for airport applications e.g. reverse osmosis, activated carbon, and oxidation (using ozone or peroxides). The main exception to date is fractional distillation which can be used in conjunction with either centralised or remote de-icing and/or stand suction cleaning, to recycle glycol based de-icing chemicals.

☐ Two biological treatment techniques have been used to successfully treat anti-icant and de-icant contaminated runoff at airports.
These are by admixture with sewage for subsequent treatment at a
sewage treatment works and land irrigation e.g. the ASG system
used at the new Munich Airport (Germany).

The use of continuous on-line monitoring equipment to monitor discharge consent compliance and provide warning of deteriorating surface water quality will increase. However, in the long term the use of
such equipment may increase an airport's chances of being prosecuted
for non-compliant surface water discharges. This is because the output
from the monitoring devices can be relayed directly to the relevant
regulatory authorities. This capability will facilitate the use of the
water quality parameter monitored as a consent parameter e.g. Total
Organic Carbon (see Appendix A). However these parameters do not
correlate well with conventional pollution parameters in airport surface water as measured by, for example, biochemical oxygen demand,
or suspended solids. Thus the use of automatic water quality monitors
could cause relatively unpolluted discharges to be non-compliant with
a consent standard specified in terms of a continuously monitorable
water quality parameter. However automatic water quality monitors do
enable real time control of drainage systems since atypical discharge
readings can be used to either trigger an emergency response and/or
activate a flow diversion system, preventing the contaminated surface
water from discharging off-site.

STRATEGIES FOR REDUCING ENVIRONMENTAL LIABILITIES ASSOCIATED WITH SURFACE WATER CONTAMINATION

Environmental legislation is becoming increasingly onerous. The
driving force behind much environmental legislation is government
accountability. This manifests itself by greater government consultation with the public. The result of this is clear to see with the
greater ease with which a member of the public can undertake law
suits for environmental offenses, the greater accountability of the regulatory watchdogs and the greater public access to environmental
information.

There are important trends in national environmental policies which
are reflected in new and proposed environmental legislation, these
include:

☐ a greater concentration on more effective implementation and enforcement of legislation, e.g. the creation of a European Environmental Protection Agency

☐ a broadening of the range of responses and instruments available to address environmental problems, e.g. the draft EC directive civil liability for damage caused by waste

☐ the increasing emphasis on preventive legislation involving greater intervention in industry, e.g. the EC directive on freedom of access to information on the environment

☐ the widening range of persons placed under environmental protection obligations.

These trends indicate that the costs and liabilities associated with surface water discharge management will continue to increase. These costs and liabilities can best be minimised by treating surface water discharge management as an integral part of the design and operation of an airport.

Insurance companies have been quick to appreciate the scale of the potential liabilities, and the costs of future environmental improvements. This is reflected in the increasing difficulty in obtaining environmental insurance. The main implication for airports of this, is with regard to the future clean up of groundwater contaminated with petroleum products. All costs associated with this activity will now be borne by the company responsible for causing the contamination, unless the cause was a 'sudden and accidental' event, which may still be covered by some insurance policies.

An airport's sewerage system will usually be wholly owned by the airport company. As such the airport company will usually be held responsible for all of the surface water discharges emanating from its sewerage system. Thus, an airport company is liable for all of its tenant's and concessionaire's contaminated discharges, be the contamination accidental, negligent or malicious.

The scale of an airport's operations is such that the provision of appropriate pollution control infrastructure will not in itself prevent the occurrence of illegal discharges. Therefore it is important that operational procedures are designed so as to minimise the risk of contaminating surface water discharges. Actions that can be taken to facilitate a process of continually reviewing and improving operational procedures include the following.

Regular monitoring Regular monitoring of discharge quality and quantity, the condition of the sewerage system, and groundwater.

Summary reports being distributed to senior managers for information and action where a non-complaint discharge is indicated.

Close liaison with regulatory authorities Close liaison with the regulatory authorities, e.g. regular monthly review meetings, helps improve the regulator's understanding of airport operations, and their willingness to advise and assist airport managers on relevant issues. In addition airport personnel gain an insight into the regulator's main concerns and a means of appearing transparent with regard to the cause of specific pollution incidents.

Education/staff awareness Staff briefings describing the environmental impact of operations, individual responsibilities and potential liabilities. Such briefings help to stimulate ideas for, and subsequently maintain, improved operating practices. Interest can be maintained by the regular publication of environmental bulletin for distribution to all airport personnel.

Close liaison on environmental matters with airport based companies Regular meetings to disseminate water quality monitoring data, information relating to recent incidents, e.g. spillages, and report progress on environmental improvement projects. etc. The aim is to cultivate collaborative environmental improvement projects in all areas of airport operations.

Reporting procedures Administrative procedures that enable incidents and/or bad practice to be reported and swiftly acted upon can be very effective at stimulating improved practices. Manchester Airport operate a Contravention/Improvement Notice System (see Appendix C) to great effect. To work effectively such a system needs two essential ingredients: the full support and involvement of senior airport managers and the full support and involvement of the regulatory authorities. Such an approach is given more credibility if clean up costs are recharged to the company responsible for a given problem.

Adoption of a Corporate Environmental Management System (BS 7750) Encompassing all environmental media and significant environmental effects.

Conclusion

The management of water discharges from an airport requires appropriate sewerage infrastructure, and the adoption of a comprehensive range of operational procedures. The preferred drainage strategy at a given airport will be that which achieves discharge consent compliance most economically, while offering the greatest protection from civil and statutory liability.

REFERENCES

1. *A Comparison of the Effects of Urea, Potassium Acetate, Calcium Magnesium Acetate, and Sodium Formate Runway De-icers on the Environment.* Transport Canada, Airports Group, Safety and Technical Services 1994, (Transport Canada Ref. TP12285E).
2. Boughey J.E., The Use of Chemical Deicants on Military Airfields in the UK, *The Public Health Engineer*, October 1984.
3. Bunce N.J., *Environmental Chemistry*, Wuerz Publishing Ltd, Canada, 1990.
4. CIBSE Guide, Volume C, Reference Data, London, 1986.
5. Chanlett E.T., *Environmental Protection*, 2nd Edition, McGraw Hill Kogakusha, Tokyo, Japan, 1979.
6. Eedy W., Salenieks S., *Environmental Impacts and Control of Glycol Based De-icer Runoff at Eight Canadian International Airports*, Beak Consultants Ltd, Ontario, 1990.
7. Featherstone R.E., Nalluri C., *Civil Engineering Hydraulics; Essential Theory with Worked Examples*, Granada Publishing, London, 1982.
8. Fisher T.J., *Trade Effluent Control, Discharges to Public Sewer*, Environmental Technology Centre Industrial Services, University of Manchester, Institute of Science and Technology, Manchester, UK, 1992.
9. Gardner H.S., Report Upon a Survey of Discharge of De-Icing Chemicals to Water Courses at Manchester International Airport, Manchester Airport Plc, June 1987.
10. Gay J., Heffcoate R., Dunn P.J., Hawkins J.E., Stormwater Contamination at Airports and Remedial Options with Particular Reference to Stansted, paper presented to ICE East Anglia Branch, July 1987.
11. *Handbook of Environmental Policies and Recommended Practices*, Transport Canada, (Transport Canada Ref. TP/21/19), Feb. 1995. Transport Canada, Airports Group, Ottawa, Ontario, Canada.
12. Harkness N., *Development of Trade Effluent Control and Charges, Waste Water Management for Industry*, IBC Technical Services Ltd, London, UK, 1990.
13. Lomas O., Avoiding Liability for Environmental Damage, *Utilities Law Review*, Autumn 1991, pp 128–133.
14. Paylor A., 1994, *Airports and the Environment*, MDIS Publications Ltd, Chichester, UK.

15. Sawyer C.N., McCarty P.L., *Chemistry for Environmental Engineering*, 3rd Edition, McGraw Hill Kogakusha, Tokyo, Japan, 1978.
16. Tebbutt T.H.Y., *Principles of Water Quality Control*, 3rd Edition, Pergamon Press, Oxford, England, 1983.
17. Vesilind P.A., *Environmental Engineering*, 3rd Edition, Butterworth-Heinemann, Newton, MA, United States, 1994.

APPENDIX A: WATER QUALITY PARAMETERS USED IN WATER POLLUTION CONTROL

Introduction

Water pollution control parameters are used to quantify polluting characteristics of water discharges. The environmental impact of water discharges can be estimated from the concentration of these parameters in a given discharge. Thus water discharge standards can be specified, using parameters that relate directly to a discharge's environmental impact.

Dissolved oxygen (DO)

Historically the main aim of water pollution abatement has been the maintenance of favourable aquatic conditions, sufficient for the growth and reproduction of normal populations of fish and other aquatic organisms. This condition requires the maintenance of dissolved oxygen levels that will support the desired aquatic life in a healthy condition at all times. The solubility of atmospheric oxygen in fresh waters is very poor. The average saturation concentration of oxygen in water is normally taken as 8 parts per million (by mass), compared to the average concentration of oxygen in air of 209,000 parts per million (by volume at sea level). In addition to this the (dynamic) viscosity of

Table 1. The minimum dissolved oxygen requirements of various types of fish

Fish/invertibrates	DO (mg/l)
Salmon/trout/mayflies	6.0–7.0
Dace/roach/caddis flies	4.0–5.0
Carp/midge larvae	3.0
Eels/fly larvae	1.2

water (0.001 Ns/m^2, at 20 degrees Celsius) is fifty times greater than air (0.00018 Ns/m^2, at 20 degrees Celsius). Thus aquatic (aerobic) organisms have to exert a considerable effort to respire and are therefore very sensitive to a pollutant's oxygen demand. The low solubility of oxygen is thus the major factor that limits the purification capacity of natural waters and necessitates the treatment of wastewaters prior to discharge to receiving waters.

Biochemical oxygen demand (BOD)

Naturally occurring aquatic micro-organisms are able to rapidly decompose 'simple' organic compounds. When aerobic microbes metabolise the organic matter they take up oxygen from the water, thus creating an oxygen demand. The biochemical oxygen demand (BOD) is a measure of a pollutant's capacity to deplete a receiving water's dissolved oxygen.

The standard BOD test measures the oxygen demand generated by a waste water sample held at a temperature of 20 degrees Celsius (assumed to be the maximum temperature in a British river), for five days (assumed to be the maximum time of flow to a sea for a British river).

Suspended solids (SS)

Suspended solids (SS) reduce light penetration into water, reducing plant growth and causing gill damage to fish. Suspended solids

Table 2. Typical BOD5 values

	BOD5 mg/l
Surface water runoff (typical UK)	5 – 50
River water (high quality)	2 – 10
Raw domestic sewage (UK)	300 – 400
Treated domestic sewage (UK)	10 – 60
Trade effluent discharges	300 – 20000
Liquid sewage sludge	10000 – 20000
Pig slurry	20000 – 30000
Silage effluent	20000 – 80000
Milk	14000
Biodegradable detergent (Aeroclean CDI)	210000
Fire fighting foam (FC 206 by 3M)	210000

are eventually deposited on the 'stream' bed suffocating macro-invertebrates and adversely affecting fish spawning success. Sources include industrial discharges, sewage treatment works, and civil engineering construction works.

Typical suspended solids values are:

☐ domestic sewage 200–500 mg/l
☐ stormwater 1–50 mg/l
☐ 'unpolluted' river water 1–20 mg/l.

Chemical oxygen demand (COD)

The chemical oxygen demand (COD) test measures the total amount of oxygen needed to completely oxidise organic matter to carbon dioxide and water. Thus the COD of an effluent indicates its potential max-imum oxygen demand. However, this test does not distinguish between biodegradable, and non-biodegradable organic matter. The major advantage of the COD test is that it can be conducted in a laboratory within four hours, instead of the five days needed for the BOD test. COD's speed of analysis selects for its use instead of BOD in many situations. The COD test is widely used to analyse the pollution strength of industrial wastes. When used in conjunction with the BOD test, the COD test is helpful in indicating toxic conditions and the presence of biologically resistant organic substances.

Nutrients – nitrogen and phosphorus

The nutrients nitrogen and phosphorus occur naturally in the aquatic environment. However, the background concentrations are usually so low that biological activity is limited. The discharge of nutrients into receiving waters can upset the delicate balance of aquatic ecosystems and cause secondary pollution effects. Nutrient over-enrichment of receiving waters is termed eutrophication.

The growth of algae (single cell plants) and bacteria in receiving waters is normally limited by the lack of nutrients. However when an increased nutrient supply is available, algae are able to rapidly assimilate it, resulting in the formation of algal blooms. Algal blooms caused by nutrient over-enrichment in receiving waters create sec-ondary pollution effects. The blooms increase the turbidity of the water, restricting light penetration into the water column. In addition, during the night when the algae cease photosynthesis, they create a net oxygen demand. The oxygen deficit generated can be sufficient to asphyxiate aquatic life, with fish kills often being the most dramatic consequence. Once the blooms begin to die off, their decomposition creates a further oxygen demand in addition to odour and turbidity nuisance.

Total carbon/total organic carbon

The total carbon (TC) or total organic carbon (TOC) content in water can be measured rapidly and continuously using automatic total carbon analysers. The analysers convert the carbon content of the water sample to carbon dioxide, the concentration of which is then measured using an infra red analyser.

The main value of TC and TOC analysis is that it can be used to continuously monitor discharges. Values of TC or TOC outside of the normal range for a given discharge can warn of deteriorating water quality. An early warning of this could initiate diagnosis and subsequent remedial action, in time to prevent a non-compliant discharge. TOC monitors are increasingly being used to automatically activate first flush diversion systems, thus enabling contaminated discharges to be automatically diverted to foul sewer and/or holding ponds.

APPENDIX B: THE FORMULATION OF SURFACE WATER DISCHARGE STANDARDS

Water discharges are passed either directly or indirectly (typically via a public sewerage systems) into receiving waters. Receiving waters include: coastal waters, estuaries, rivers, streams and groundwater.

Regulatory authorities will normally issue discharge consents to the person or company that owns the land from which the discharge emanates. A discharge consent for a given site will usually relate to a given discharge at a point where it outfalls directly into a receiving water or a public sewer. A water discharge is only legitimate if the outfall or public sewer connection is approved and if the discharge complies with limits on flow rate and composition specified by regulatory authorities.

One or more regulatory authorities will be responsible for specifying the discharge consent standards, and monitoring compliance. In specifying the constraints for a given discharge consent the regulatory authority is guided by:

☐ international agreements
☐ national legislation

☐ its own bylaws
☐ the assimilative capacity of the receiving water
☐ the water quality objective of the receiving water.

The assimilative capacity of a receiving water reflects its ability to render a discharges' contaminants innocuous. Receiving waters in ascending order of assimilative capacity include coastal waters, estuary, rivers, stream and groundwater. Thus coastal waters are typically better able to ameliorate the effects of polluted discharges than groundwaters. The regulatory authority will set water quality objectives for receiving waters, and then regularly reduce the discharge consent levels until the objectives are achieved. A discharge consent is thus site-specific and states both the maximum pollution strength and maximum flow rate of a particular discharge. An example of a discharge consents water quality specification is:

maximum permissible concentrations: 10 mg/l BOD
50 mg/l SS
2 mg/l ammoniacal nitrogen
no visible oils.

This discharge consent standard is presently (1995) being imposed on the stormwater discharges from Manchester Airport, and reflects the increasingly stringent discharge standards being sought by regulatory authorities.

APPENDIX C: CONTRAVENTION/IMPROVEMENT NOTICE
Manchester Airport Plc 1991

Serial No. 0331

CONTRAVENTION / IMPROVEMENT NOTICE

Date .... Issue To (Company / Dept) ....

Address ....

....

By .... Signed .... Ext ....

Problem ....

....

.... Sheet Attached ....

Location of Problem ....

....

Unless remedied this is likely to lead to contravention of /
This is in contravention of ....

....

Regulator(s) ....

******* THIS IS ALSO IN BREACH OF CONTRACT / TENANCY *******

Remedial Action must be taken immediately by you to resolve the above and ought to include:

....

....

.... Sheet Attached ....

A MANCHESTER AIRPORT PLC REPRESENTATIVE WILL RE-VISIT AND MAY BE ACCOMPANIED BY THE REGULATOR WHO MAY TAKE APPROPRIATE ACTION.

MANCHESTER AIRPORT PLC IS TAKING / MAY ALSO TAKE ACTION INCLUDING:
TERMINATION OF CONTRACT, IMMEDIATE EJECTION FROM SITE, RECOVERY OF COSTS FOR REPARATION, DAMAGES, LEGAL ACTION, DIRECT INVOLVEMENT OF THE REGULATOR, CLOSING THE ACTIVITY OR YOUR OPERATION DOWN.

ADVICE ON THIS ISSUE MAY BE AVAILABLE FROM M.A. PLC TEL 061-489 3776 OR INTERNAL EXTENSION 3776. INABILITY TO ACCESS ADVICE FROM M.A. PLC WILL NOT BE ACCEPTED AS REASON FOR NON COMPLIANCE.

Signed...................... CHIEF EXECUTIVE MANCHESTER AIRPORT PLC

I understand the implications of the above and accept this prohibition / improvement order on behalf of:

.... Title ....

Signature .... Date ....

Surface water at Stockholm-Arlanda Airport

PETER O VIKSTROM, Swedish Civil Aviation Administration, Stockholm-Arlanda Airport, Stockholm 1, S-19045 Stockholm Arlanda, Sweden

This paper describes some of the steps already taken at the airport to reduce the discharge of pollutants in storm water into local watercourses. It briefly discusses measures being planned in connection with the addition of a new runway at the airport. Based on a knowledge of the water quality situation today, it examines what further measures are necessary to minimise the airport's environmental impact in this area. The pollutants discussed are primarily organic substances, petroleum and heavy metals.

DESCRIPTION OF SURROUNDINGS

Stockholm-Arlanda Airport is located in a small drainage area, totalling approximately 8000 hectares. This area is dominated by agricultural and forest land but also includes a small community with about 32,000 inhabitants. The airport was previously the dominant discharge source in the area, since in principle there are no other significant industrial operations. The different types of land in the drainage area are estimated in Table 1.

The size of the airport, inside the perimeter fence, is 630 ha.

Table 1. Drainage area by type of land (1 ha ~ 2.5 acres)

Type of land	Size
Forest	3900 ha
Farmland	3060 ha
Paved surfaces	950 ha
Wetlands	70 ha
Lakes	50 ha

The airport is drained by two watercourses: Kättstabäcken (Kättsta Creek), which runs west of the airport, and Märstaån (the Märsta River), which runs south of the airport. These two watercourses meet about 2 km south-west of the airport. The average water discharge at this point is about 200 litres per second, while the average discharge from the entire drainage area is about 500 l/s. Because of these relatively small water flows, the discharge of pollutants in storm water from the airport has a major impact in downstream watercourses.

The geological conditions in the area are highly varied. The airport is situated in an area where nearly every imaginable kind of soil is represented. Runway 01/19 rests on a combination of moraine, clay and rock. Runway 08/26 rests on moraine, clay, sand and fluvio-glacial gravel. The terminal and operations area is located on moraine and rock. The various kinds of soil have been utilised to drain storm water from runways in such a way that it both infiltrates the ground and is discharged as surface run-off. In those sites where the ground has high hydraulic conductivity (sand and gravel), there are no conventional storm water drains. Instead the water infiltrates directly into the ground. In other parts of the runway system with relatively dense kinds of soil, the storm water is piped away from the airport area.

One way of indicating the scale of airport operations is to report annual flow volumes of various chemicals. Table 2 shows the quantities of certain key chemicals used at the airport during 1994.[1]

The following example shows the importance of recovering these chemicals. If all de-icing fluid used at the airport – 1270 tonnes of propylene glycol – were permitted to run straight into the local watercourses, the average level of organic substances in the drainage area would be 40 mg of dissolved organic carbon (DOC) per litre. This would signify a 400 per cent increase from the natural background level.

Table 2: Use of certain key chemicals, 1994 (1 tonne = 1000 kg or 2204 lb)

Chemical	Specification	1994 quantity
Jet fuel	JET A-1	530,000 m^3
De-icing fluid type-I (100%)	Propyleneglycol	1100 t
Anti-icing fluid type-II (100%)	Propyleneglycol	170 t
Runway de-icing agent	Potassium acetate	300 t
Fire-fighting fluids	Detergent, AFFF	15 m^3
Fire-fighting powder	$K_2SO_4 + CaCO_3$	15 t

COLLECTION OF DE-ICING FLUID

The system for collecting spilled de-icing fluid is largely based on previously laid storm water drains and pipes. These were renovated so that valves can be used to shift between winter and summer operation. In areas where de-icing of aircraft takes place in winter, there is a separate storm drainage system so that the ordinary surface water is not affected. Glycol-polluted surface water is gravity-piped to a number of pump stations, from which the water is pumped into a balancing pond. When snow falls, it is collected separately at de-icing sites and placed in a special snow dump, which is sealed with a geomembrane to prevent the infiltration of pollutants into the groundwater. To facilitate this separation, a red dye is added to the de-icing fluid. This makes the glycol-polluted snow easier to see. When the snow melts, the meltwater runs from the dump to the above-mentioned balancing pond.

The water that has been collected in the balancing pond goes through a

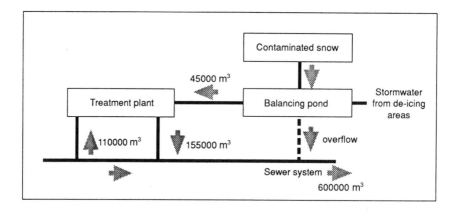

Fig. 1. System for collecting and breaking down de-icing fluid

Table 3: Sizes of units in system

Unit of system	Size
De-icing areas	160,000 m^2
Balancing pond	9000 m^3
Polluted snow dump	40,000 m^3
Treatment plant	3000 m^3

treatment plant to reduce the level of organic substances. To raise the temperature of the fluid and add certain nutritive salts, sanitary waste water from the airport is used. Some of this waste water is pumped into the treatment plant. Glycol-polluted surface water is added to it by pumping from the balancing pond after regular monitoring of the level of organic substances, based on chemical oxygen demand (COD). The treatment plant uses an activated sludge method. It is furnished with aeration and decanting equipment and operates as a batch biological reactor. The plant is designed for COD reduction of about 60 per cent. After treatment, the water is pumped back into the waste water network for further treatment at the municipal sewage plant.

Table 3 gives the size of the units in Fig. 1.

The operating permit that has been granted to the airport under the provisions of the Environmental Protection Act imposes conditions concerning the quantity of COD that may be discharged into the waste water network and the quantity of de-icing glycol that may be discharged into the local watercourses. In the latter case, the permit states that 10 per cent of de-icing glycol used during a de-icing season may be discharged in storm water into local watercourses. Over the past three winters, such discharges have totalled around 5 per cent of the de-icing fluid used.

COLLECTION OF FIRE-FIGHTING FLUIDS

Water containing residues from fire-extinguishing and fire-ignition materials is often very toxic to waterborne organisms. This has been confirmed in tests of waste water containing chemicals from fire-fighter training at the airport (Microtox test). Tests of the degradability of the same waste water have shown that the water is, in fact, easily degradable but that there is a residue of organic substances that is very difficult to break down.[2] As a result, water from fire-fighter training activities is unsuitable for untreated discharge into watercourses.

The fire-fighter training areas at Arlanda were renovated some years ago and equipped with a new drainage system. These areas consist of five detached concrete and asphalt zones from which polluted surface water is piped to a 500 m^3 collecting pond. In addition, there is a fire pond that is drained manually into a cast concrete tank. In recent years, the polluted water in the collection pond has been hauled away for destruction, but this process has proved very costly. Consequently, during 1994 the airport began an experimental project for treatment of this water. The experiments are performed in situ on a pilot scale. The method employed is chemical precipitation, combined with flotation type separation plus final filtration through activated charcoal. The treatment has proved very effective. Removal of petroleum has been >99 per cent, and removal of organic substances >95 per cent (COD). At the end of 1995 these experiments will be evaluated in detail.

NEW MAINTENANCE AND CARE INSTRUCTIONS FOR OIL SEPARATORS

A very simple measure that has reduced petroleum discharges is the introduction of an effective maintenance program for the oil separators at the airport. About ten gravimetric oil separators are connected to the storm water system, and run-off from areas where aircraft are parked and fuelled is piped to these separators.

This program contains instructions on frequency of maintenance, monitoring of alarm systems, oil level measurements, emptying procedures and record-keeping. All oil separators are inspected by the airport's heating, water and sanitation department once a month.

STORM WATER FROM THE PLANNED NEW RUNWAY A6

In April 1993, Stockholm-Arlanda Airport received final permission under the provisions of the Environmental Protection Act to extend the airport and operate it with three runways. This permission was granted by the National Licensing Board for Environmental Protection. According to plans, the third runway will go into service in 1999. Storm water drainage from the new runway is being planned in such a way that surrounding watercourses will not receive de-icing agents or other chemicals such as petroleum hydrocarbons that are spilled on the runway. The runway will be equipped with a 10 m wide shoulder with storm water drains at their low points. Run-off will be gravity-piped to a balancing pond designed to hold about 20,000 m^3. This is large enough for rain of 16 hours duration, recurring every second year. Parts of this pond will function as a sand trap and oil separator. Land has also been reserved for a supplementary treatment stage, in case storm water treatment should become necessary.

Parts of the new runway system will be located on a fluvio-glacial deposit that contains a substantial groundwater resource. In keeping with the Environmental Protection Act, this groundwater will be shielded against pollution from airport operations. This will be achieved by laying a geomembrane at a depth of about 1 m next to the runway system, in areas where the ground consists of sand and gravel. Run-off from this geomembrane will also be piped to the balancing pond.

The need for a protective geomembrane has recently been questioned, especially because the airport has changed its runway de-icer to an acetate-based agent. However, the philosophy of the airport is that this geomembrane should be viewed in a hundred-year perspective, given the risk of sizeable discharges of fire extinguishing chemicals and petroleum products in case of leaks or accidents.

MEASURABLE IMPROVEMENTS

The quality of receiving waters around the airport has gradually improved in recent years. One 1993 study of the benthic community (bottom-dwelling fauna) in local watercourses showed unequivocal improvements in water quality conditions.[3] Table 4 shows the change in the number of species at four sites in these watercourses between 1988 and 1993. It demonstrated an increase in biodiversity, with a fauna closely resembling the supposed ideal.

During the same period, there were especially significant reductions in organic substance and nitrogen levels as a consequence of the measures undertaken at the airport. The average DOC count in the receiving waters declined from about 25–50 mg/l DOC to about 15 mg/l. Today the total nitrogen level is down to around 2 mg N/l, compared to about 5–10 mg/l in the late 1980s.

Table 4. Number of species (taxons) in local water-courses, 1988 and 1993

Site	Number of species		
	1988	1993	Change
UP 78	6	22	+16
UP 79	4	25	+21
UP 80	3	12	+9
UP 82	1	6	+5

DISCUSSION

What remaining problems exist with regard to discharge of pollutants in storm water? It may be worth mentioning a number of potential environmental risks or problems at Stockholm-Arlanda.

Sediment samples taken in the vicinity of the airport have shown higher levels of various metals such as copper, cadmium and zinc. Compared to the Swedish National Environmental Protection Agency's criteria for measuring the metal content in sediments, the level of cadmium at some sites is classified as high, >2 mg/kg (dry-weight basis). It is not surprising that cadmium is found in these sediments, since we know that waste water from hangars becomes polluted with this metal when aircraft are washed. The same phenomenon can therefore be suspected when aircraft are subjected to precipitation. One conceivable solution would be for the airport to sediment these metals in ponds or separate them in sand filters, before storm water is allowed to flow out of the area.

A crash or a major leak of jet fuel on the existing runways would cause major damage to the receiving waters. Since the storm water from these runways does not pass through any separation systems, in such a situation we must rely entirely on clean-up measures. The success of a clean-up operation will be determined by how prepared the airport is for this type of incident. An emergency plan that includes clear instructions concerning clean-ups would probably be one way of reducing the risk of environmental damage.

Acetate-based runway de-icing agents have been in the market for some time. The airport has used these agents since 1992 and has seen how they have improved the environment. Various experiments in the break-down of acetate in the ground, for example in Norway,[4] have shown that in sandy soils, the natural degradation process is effective up to levels of around 5 g KAc/l (equivalent to about 1300 mg/l DOC) but that higher levels of acetate may lead to salt stress and thereby inhibit the break-down of acetate. There are also lingering questions about the long-term impact of this chemical.

In a cold watercourse, break-down is likely to proceed more slowly than in the soil. At Stockholm-Arlanda, surface water from parts of the runway system runs straight into the local watercourses. Although acetate contains a far smaller quantity of organic substances than does glycol, acetate does cause some pollution. This pollution, along with diffuse discharges of aircraft de-icing fluids, together create a situation where treatment measures should be considered. The average level of organic substances in the storm water from runways and ramps is around 25–30 mg/l DOC during the year. A reduction is desirable. During 1995 the airport will start a project aimed at identifying low-cost natural methods of reducing organic substance levels.

REFERENCES

1. Vikstrom P (1994) 'Årsrapport av dagvattenutsläpp och kemika-lieanvändning vid Stockholm-Arlanda flygplats (*Yearly Report on Storm Water Discharges and Chemical Use at Stockholm-Arlanda Airport*)', Luftfartsverket, Stockholm-Arlanda.
2. VVL (1993) 'Karakterisering av prov från brandövningsplatsen på Arlanda (*Characterization of Sample from the Fire Training Site at Arlanda*)', Laboratorierapport, Stockholm.
3. Lingdell P.E, Engblom. E (1993) 'Kan smådjur leva i Märstaåns vattensystem ? (*Can small animals live in the Märsta River Water System?*)', Limnodata HB, Skinnskatteberg.
4. French H.K, Roseth R, Englund J-O, Meyer K-F, Swensen B, Hongve D (1994) 'Rensekapasitet i jord - Avisningskjemikalier; transport og nedbrytning i jord (*Purification Capacity of Soil – De-icing Chemicals, Transport and Degradation in Soil*)', Norges landbrukshøgskole för Oslo Hovedflyplass Gardemoen.

Airfield groundwater contamination – an overview

ROGER JOHNSON, Camp Dresser and McKee Inc., Ten Cambridge Center, Cambridge, Massachusetts 02142, USA and NADINE TUNSTALL PEDOE, Scott Wilson CDM, Bayheath House, Rose Hill West, Chesterfield, Derbyshire S40 1JF, UK

This paper focuses on the following topics

☐ the type of contaminant found at airports and how these exist in groundwater
☐ the sources of groundwater contamination at airports.

It also discusses some of the regulatory requirements facing the United States, a few case studies, and some remedial technologies for contaminated groundwater.

TYPES OF CONTAMINANT

The following is a list of the most common types of groundwater contaminant found at airports. These mostly consist of petroleum hydrocarbons and fuels, including jet fuel, and diesel fuel. The number of carbon atoms influences the behaviour of the contaminant.

☐ petroleum (C4 - C12)
☐ jet fuels (C9 - C16)
☐ diesel fuels (C9 - C22)
☐ chlorinated solvents
☐ deicing/anti-icing compounds
　– 2 glycols
　– urea.

Cx = number of carbon atoms in the hydrocarbon compounds.

Chlorinated compounds (i.e. solvents and degreasers) are another common group of contaminants found at airports. While chlorinated compounds do not constitute a significant quantity of contaminants at airports, the experience in the USA is that they are one of the most problematic. This is because most chlorinated compounds are carcinogenic, and are extremely mobile in the subsurface environment. Most regulatory agencies in the USA responsible for groundwater quality routinely monitor for the presence of chlorinated compounds, and become concerned when they are discovered in, or near, drinking water supplies.

Another common contaminant found at northern airports are those resulting from the use of deicing and anti-icing compounds (glycols and ureas). Although deicing and anti-icing compounds more typically impact on surface waters, and there are no drinking water quality standards for glycols in the USA, these compounds have been found to be a significant problem in the subsurface within the USA. The most significant impact of these compounds appears to be the odours associated with biodegradation. This has been a problem at Milwuakee International Airport, where there is a shallow groundwater regime.

CONTAMINANT PHASES

There are a number of compounds that can contribute to groundwater contamination, and the way these contaminants enter, and behave in, groundwater varies significantly. Fig. 1 demonstrates the four most common phases of petroleum contamination associated with a leaking underground storage tank.

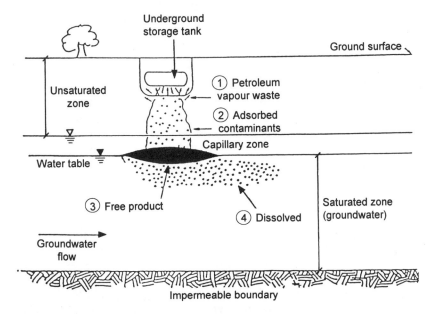

Fig. 1. The subsurface environment and four phases of contamination

The vapour phase

Volatile organic compounds, like petroleum, can exist as vapours within the pore spaces in soil and can migrate a significant distance from the source by several different mechanisms, of which the most significant is diffusion. The vapours can be later dissolved in infiltrating water such as rainwater and contaminate groundwater.

The sorbed product phase

The contaminant exists as free product bound to soil particles, by a process termed 'residual saturation'. Contaminants in this phase can also become dissolved into infiltrating water such as rainwater, which has the potential to further contaminate groundwater.

Contaminants in both the vapour and the sorbed phase must be addressed when planning any remediation of contaminated groundwater. Both can act as continued sources of contamination prolonging remediation efforts.

The 'free product' phase

This generally occurs along the capillary fringe. If a spill occurs which exceeds the residual saturation capacity of the soil, free product will migrate down to the soil/water interface, or the top of the saturated zone. It should be noted here that Fig. 1 represents petroleum or other volatile organic and/or semi-volatile organic compounds, that have a specific gravity less than 1 (i.e. lighter than water). Some compounds such as chlorinated solvents have a specific gravity greater than 1. These compounds can migrate as free product past the soil/water interface and exist as free product within the saturated zone. The compounds will stop migrating only when encountering a natural barrier such as a clay lens, or when all the material becomes sorbed onto soil particles within the soil and water column.

The dissolved phase

Contaminants will dissolve into water at a rate based on the solubility of the material. Dissolved compounds can then migrate in groundwater and potentially contaminate off-site drinking water supplies.

Figure 2(a) and (b) illustrate some of the different characteristics of groundwater contamination. Fig. 2(a) shows a situation in which the contaminant is in direct contact with the water table, resulting from,

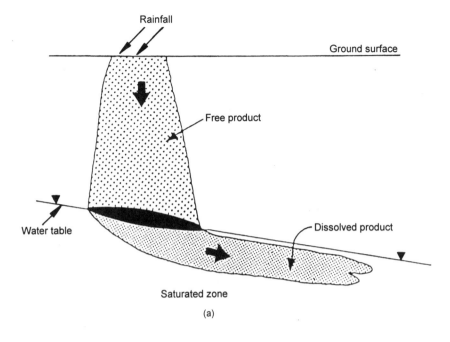

(a)

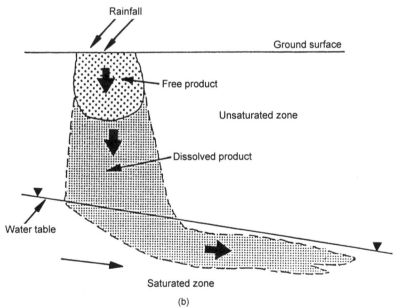

(b)

Fig. 2. Contaminant plumes showing methods by which groundwater can be contaminated: (a) contaminant in direct contact with the water table; (b) ground-water contamination resulting from solution of contaminant in percolating recharge water

for example, a hydrant fuel leak which has the potential for the release of a significant amount of product. This product exists from the surface down to the water table. In this situation, infiltrating water acts as a transport mechanism to drive this product down to the water table. The dark area above the water table area represents free product. The grey area represents dissolved product, which migrates in the direction of groundwater flow.

Figure 2(b) shows a more typical plume, in which groundwater contamination results from contaminants dissolving in infiltrating water. This plume is characterised by sorbed product which does not migrate as free product to the water table. In this instance, infiltrating waters such as rainwater, leaking stormwater drains, runway pavement washwater, leaking utility lines, etc. will act as a transport mechanism to drive the contaminants down to the water table where they can migrate. This type of plume is more difficult to detect than one in which free product is present at the water table.

CONTAMINANT SOURCES

Bulk fuel storage facilities

The largest source of contamination at airports is bulk fuel storage facilities. At most existing airports, above-ground bulk storage tanks were installed prior to the development of environmental safeguards. For example, in the USA, most above-ground tanks are single bottom tanks which may be situated in unlined containment structures. These structures provide little to no protection should a leak occur in the tank.

Hydrant fuel systems

These are another significant source of groundwater contamination. Hydrant fuelling is performed by connecting a hydrant cart to a hydrant fueling system. Fuel is pumped through a subsurface pipeline to a hydrant pit where the connection is made. In the past a small amount of fuel spillages would occur whenever the cart was connected or disconnected. Pits would fill with fuel which could be a nuisance. To relieve this situation, many hydrant pits were constructed with a gravel base to facilitate drainage of fuel out of the pit. This practice has resulted in a significant amount of groundwater and soil contamination around hydrant fueling systems. For example, Terminal Two at the Los Angeles International Airport (LAX) was recently expanded. During the expansion project, it became necessary to cut through the existing ramp and relocate several of the hydrant lines. As the concrete was removed and trenches were dug, the trenches filled with jet fuel. It was determined that the soil under the ramp was virtually saturated in some areas due to the historic leaks from the hydrant system. The groundwater is 27 m below the ground surface, and jet fuel has been found in some areas to have saturated to a depth of 30 m, with 3.5 m of free product floating on the groundwater surface.

Leaking underground storage tanks (USTs)

USTs represent one of the largest single sources of groundwater contamination in the USA, which has embarked on a significant programme to remove old leaking USTs and require leak detection and spill protection on all new USTs. However, USTs still represent a significant source of contamination. The USA Environmental Protection Agency (USEPA) estimates that fewer than half of the estimated two-million USTs in the USA have been removed and/or upgraded.

Maintenance activities

These activities include, but are not limited to, aircraft and ground vehicle maintenance, printing and stripping and facilities maintenance. The real problem associated with these activities is the storage, handling, and disposal of hazardous materials used during the carrying out of maintenance activities.

CASE STUDIES IN AIRFIELD GROUNDWATER CONTAMINATION

Figure 3(a), (b), and (c) show the old Terminal Building at Pittsburgh International Airport in Pennsylvania, which is currently being decommissioned as a new terminal has been constructed.

The lines adjacent to the Terminal represent the hydrant fuel system, which as Fig. 3 shows, has contributed to soil and/or groundwater contamination. One additional problem is that not only do these hydrant lines represent sources of contamination, they also represent a significant migration pathway. Leaks that originate from a pit or other section of pipe can migrate along the backfill of the hydrant line a significant distance from the site of the leak. This is especially true of hydrant systems constructed in areas with naturally occurring low-permeability soils.

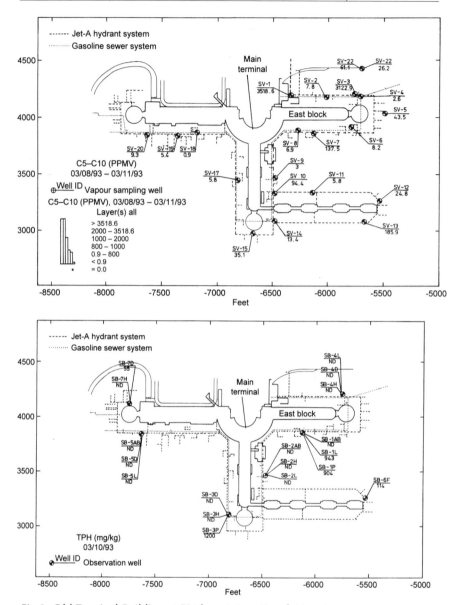

Fig 3. Old Terminal Building at Pittsburg International Airport

Figure 4 illustrates groundwater contamination at LAX. Although groundwater contamination is ubiquitous at major airports across the USA, this contamination is generally undetected until an investigation is triggered by an incident. For example, at LAX, groundwater contamination was discovered as part of a routine property transfer

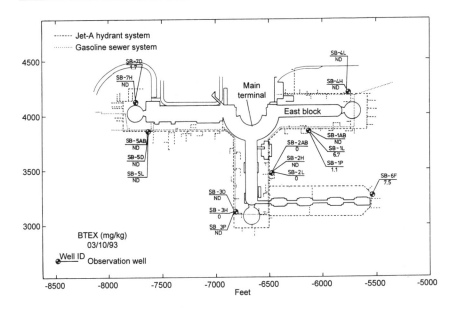

Fig. 3. (continued)

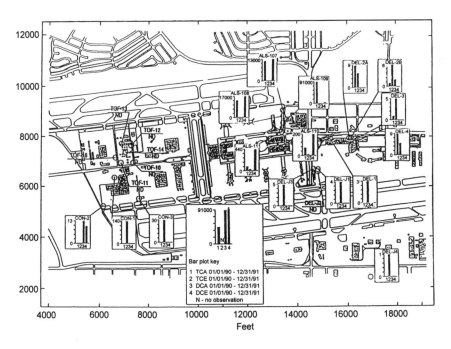

Fig. 4. Los Angeles International Airport (LAX) Plan showing fingerprints of chlorinated solvents

assessment. In the USA the purchaser of contaminated property can be held liable for contamination.

In order to protect themselves from this liability, property purchasers must demonstrate that they have exercised 'due diligence' in investigating the property. This is accomplished through the conduct of an environmental assessment.

In about 1984, the oil companies who had operated the fuel storage, transfer, and hydrant fuel system at LAX sold the systems to a fuelling consortium. The consortium conducted an environmental assessment of the property and have discovered about 3 m of free product on the water table. This product may have gone undetected for several more years had the property transfer process not been initiated.

In Arizona, the Phoenix Sky Harbour International Airport sits in the middle of a large area of groundwater contamination. This area has been designated a state-lead 'Superfund' site. Chlorinated solvent contamination has been discovered in agricultural wells directly down gradient of the airport. As a result of this contamination, the airport has been forced to initiate a programme to investigate potential sources of contamination at the airport.

In many cases, groundwater contamination has been discovered only as airports close.

☐ Denver Stapleton International Airport, in Denver, Colorado has the same problem. Thousands of litres of jet fuel have been removed from beneath the apron at the airport. The source appears to be nothing more than historic spills and leaks associated with the fuel hydrant system.
☐ Miami International Airport has an identical problem, with a shallow aquifer system, and 1.4 m of floating free product.
☐ Ontario International Airport in California has a unique problem. Chlorinated solvent contamination has been discovered in agricultural wells down gradient of the airport. An unlined surface water drainage channel runs directly through the airport, which has been identified as one of the sources of this contamination.

REGULATORY REQUIREMENTS

Regulatory requirements in the USA vary by state, county and city. For example, in California, there are 64 agencies with jurisdiction over USTs. These agencies are increasingly targeting airport authorities instead of the tenants, who in most cases are responsible for the problems. Just as businesses are facing a future with limited resources, so too are regulatory agencies. These agencies are finding it easier, at least in the USA, to focus on a single authority rather than 50 or 60 tenants. Airports, as a result, are becoming less a regulated community and more an extension of the regulatory agencies with respect to dealing with tenants and environmental compliance with the regulations.

Airports can therefore influence agency direction. The following two case studies highlight two airports in California. Both airports have similar problems, and are regulated by the same agency. Orange County's John Wayne International Airport has tenant-caused jet fuel and chlorinated solvent contamination of groundwater. This contamination exists in water that is non-potable due to naturally occurring conditions. As previously discussed, Ontario International Airport in California has been found to be a source of chlorinated solvent contamination of down-gradient agricultural wells.

John Wayne Airport

The groundwater problem at John Wayne Airport was discovered through a construction project. A depressed roadway was constructed on the airport which required dewatering during construction. The water produced during the dewatering activities was found to be contaminated and over US$1 million was spent in treating this water so it could be subsequently discharged to a storm drain. Following completion, the regulatory agency issued a clean-up and abatement order to the airport. The Agency's opinion was that, while the tenants were primarily responsible for the contamination, as the property owner, the Authority has ultimate responsibility. In addition, the regulators indicated that it was easier for them to deal with one party, i.e. the Airport Authority than to deal with multiple tenants. The Authority is now being forced to conduct a multi-million USA dollar investigation into the causes of the contamination and the sources.

The airport may be forced to file lawsuits to recover costs. In some cases, the tenants have abandoned the contaminated leaseholds, and the airport has to clean up the property before it can be re-leased to another tenant.

Ontario International Airport

This airport presents a similar situation but a completely different outcome. In 1988, the City of Los Angeles, the owner of Ontario

International Airport, implemented a comprehensive environmental management programme covering all airports owned by the city. This programme began with an extensive environmental audit of all airport owned property, including the property leased by tenants, and included a review of all applicable regulations and identification of all regulatory agencies with jurisdiction over the airports. Under the programme, a task force comprising the airport and regulatory agencies was formed. As a result of this relationship with the regulators, when the groundwater contamination was found down-gradient of the airport, the Authority was able to deflect the agencies' attention to the tenants. The airport is now directing the tenant activities in partnership with the agencies. In come cases, the Authority has successfully convinced the agencies to trace tenants who left the airport in the late 1950s. The important message here is that it is the tenants that are being held responsible for the clean-up, not the airport. It is important to establish a relationship with the regulatory agencies before a problem is discovered. This relationship will allow the airport to deal effectively in a partnering relationship if problems are then discovered later on.

Development of clean-up standards for groundwater, especially groundwater that has become polluted, is currently a major issue in the USA. Historically, USA regulatory agencies have required industries to attempt to remove all contamination to below drinking water quality standards. For example, the state of California has an official 'non-degradation' policy. Simply stated, this policy says 'if you want to pollute, you clean it up'. The problem with this policy is that the regulations allow no consideration of the ultimate use of the water. Requiring industries to attempt to restore an aquifer to drinking water standards when the water is non-potable due to naturally occurring compounds such as metals, total dissolved solids, etc is not reasonable. In addition, after 20 years of active groundwater remediation projects, the USA has discovered that it is technically impracticable and impossible to restore an aquifer to drinking water standards in most cases.

As a result of the above, USA regulators are leaning towards health-risk based clean-up standards. The USEPA has recently issued a technical memorandum on implementing what is called the 'technical impracticability' argument for not cleaning up groundwater. Different states have implemented risk-based standards for the clean-up of soil and groundwater.

It is important that airport authorities look at this issue from two perspectives: that of the authority and that of the landlord.

From the perspective of the authority it may be possible to argue for 'risk-based', more lenient clean-up goals citing the experience of the USA, if required by regulatory agencies to implement a remediation programme.

However, the landlord must be aware of what kind of goals the tenants may be negotiating. Most negotiated goals will require some form of deed restriction which mandates a review by the regulatory agencies before any future change in land use. As an Authority this is generally not acceptable as it may defer the expense of clean-up from the tenant back to them, if and when the tenant leaves.

REMEDIAL TECHNOLOGIES

As previously stated, in most cases, complete restoration of contaminated groundwater is, for the most part, technically impracticable. One exception to this may be a permeable aquifer contaminated with a volatile, biodegradable material. However, groundwater remediation must still be initiated, if free product exists or if the contamination is impacting drinking water supplies. The following briefly present some remedial technologies that have been effective in the USA.

Containment

This ensures that the groundwater contamination does not spread. The use of hydraulic systems include groundwater extraction wells and trenches, which can be pumped and then provide a hydraulic barrier. Slurry walls containing concrete or clay slurry followed by pumping can be used in shallow groundwater areas.

Free product removal

Figure 5 shows a single pump recovery system for the recovery of free product. This system is not designed for dissolved product. A single pump is placed in a well containing free product. In order for this system to work, the well must be completed above and below the water/product interface. The pump intake is set slightly below the product level and pumps both water and product. The problem with this system, is that the water that is removed becomes contaminated through the process and must be treated.

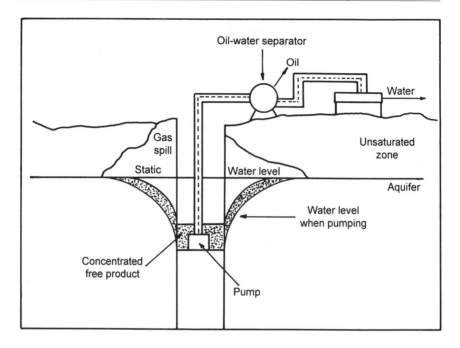

Fig. 5. Single pump recovery system

Figure 6 illustrates a Dual Pump Recovery System. This system is also designed to recover free product. However, this system is designed to substantially reduce the amount of water that becomes contaminated by using the single pump system. In this system, one pump is set below the water table and this pumps clean water. The purpose of this pump is to depress the water table allowing product to flow into the well. A second 'skimmer' pump is set higher to skim off the product. As the product levels decrease, the skimmer pump must be moved down. In addition to reducing the amount of water requiring treatment, this system allows product to be recovered in a clean enough condition to allow some 'salvage' value from the recovered product to offset the cost of the remediation.

The most promising technology for airports is in-situ bioremediation (Fig. 7), which has been pioneered by the USA Air Force Centre for Environmental Excellence on approximately 200 air force bases. This technology involves injecting oxygen, nutrients, and in some cases biological cultures into the groundwater to induce and/or enhance biological activity.

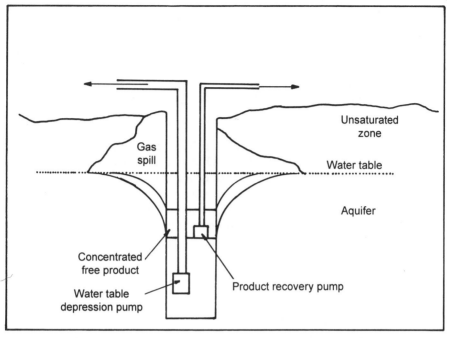

Fig. 6. Dual pump recovery system

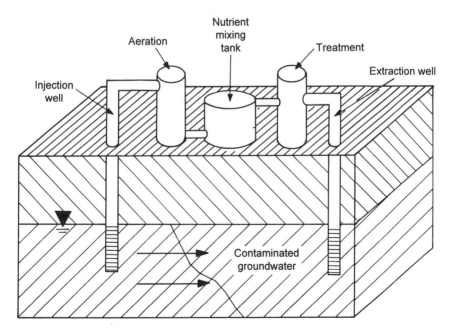

Fig. 7. In-situ bioremediation

The USEPA is beginning to recognise a technology they call 'indigenous biodegradation', which is nothing more than a term for doing nothing. Through a significant amount of research, the USEPA has concluded that in many cases dealing with biodegradable materials such as benzene, ethyl-benzene, toluene, and xylene, natural or indigenous biodegradation occurs at a rate faster than a plume migrates through the aquifer.

This argument has been used successfully at a number of sites, in the experience of one of the authors. One example was at a terminal facility on Long Island, New York. In order to contain a plume on this site, a system capable of pumping in excess of 50 million litres a day would have been required. The consultants were able to convince the regulators that the dissolved gasoline plume was biodegrading faster than the groundwater was flowing. This has resulted in a saving of millions of dollars.

The last technology was also developed and named by the Air Force. Bio-slurping (Fig. 8), is actually a combination of technologies for the remediation of both free product as well as dissolved phase product in the soil and in the groundwater.

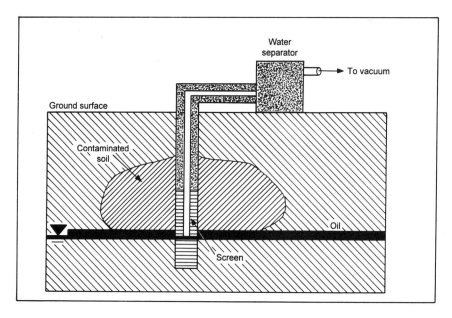

Fig. 8. Bio-slurping

Bio-slurping involves the completion of a well used for the recovery of free product as well as for the injection and/or extraction of oxygen into the soil and/or groundwater. Bio-slurping is a dual phase technology designed to recover free product, by pumping, and as a vapour phase or air sparging well. The well is screened above the water table, and the free product pumped out. The air is pulled out in a vacuum to vapour extract the soils above the saturated zone. Alternatively the air can be injected for air sparging or bioventing techniques, where the soils are aerated to enhance biodegradation of contaminants below the water table. This technique has been found to be the most efficient, cost effective and the fastest acting of the remedial technologies.

Groundwater protection measures at the new Munich Airport

HELMUT HOFSTETTER, Department of Civil Engineering and Water Management, Flughafen München GmbH, Nordallee 43, D-85326 Munich, Germany

GENERAL INTRODUCTION

During the planning, construction and operation of Munich Airport, water conservation issues played an important role because the airport is located in a sensitive area from the point of view of water conservation.

This report looks at the planning, construction and operation work to examine how water management problems were solved, in particular the problem of pollution control.

The new Munich Airport is located 28.5 kilometres to the north-east of Munich. It is served by the city's S-Bahn transit railway and links up to the national network of autobahns via the A 92, which connects Munich and Deggendorf. The airport has two runways, each 4000 m long, built parallel to one another with a separation of 2300 m, and which extend from east to west. Between the two runways are the main terminal, the west apron, which has 59 aircraft stands, and the northern and southern development zones in which air cargo, aircraft maintenance and administration buildings, operational facilities, workshops, the supply centre, and catering facilities were constructed. The airport also has other areas held in reserve for future expansion. The total airport site has an area of roughly 1500 hectares.

The airport's planning and construction phase was a lengthy and arduous process spanning 30 years, with the following key milestones:

- [ ] March 1963: the 'Munich Airport Site Commission' is appointed.
- [ ] October 1966: the Bavarian Government begins the regional planning procedure for a new Munich Airport at the proposed Hofolding Forest site.

Fig. 1. Location of Munich Airport

☐ October 1967: the regional planning procedure is expanded to include the site Erding-North.

☐ August 1969: the Munich Airport Authority applies for the approval required by aviation law for the new airport Erding-North/Freising.

☐ May 1974: the Bavarian State Ministry for Economic Affairs and Transportation grants operational authority for the construction of the new Munich Airport at the Erding-North/Freising site.

☐ June 1974: the Munich Airport Authority applies for the second approval procedure to be carried out; the designation order from the Government of Upper Bavaria.

☐ July 1979: the Bavarian district government finalises the public planning approval decision for the new Munich Airport and orders 'immediate implementation'.

☐ November 1980: construction work on the new Munich Airport begins.

☐ April 1981: the Bavarian Administrative Court in Munich halts construction, reversing the district government's 'immediate implementation' order for the new airport.

☐ March 1985: the Bavarian Administration Court re-affirms the 'immediate implementation' order. Construction resumes.

☐ December 1986: the Federal Administrative Court in Berlin, the third and final appeals level, ultimately declares the Munich plans legal.

☐ September 1989: topping-out ceremony for the new airport Terminal is held.

☐ 17 May 1992: the new Munich Airport begins operations.

GEOLOGICAL AND HYDROGEOLOGICAL CHARACTERISTICS

The new Munich Airport is located on the northern rim of the Munich gravel plain at 448 m above sea-level. The Munich gravel plain, which extends from the foothills of the Alps to the town of Freising, developed through glacial deposition during the Riss and the Würm (roughly 50,000–10,000 years ago), in other words during the Quaternary Period. The gravel deposits reduce in depth from approximately 90 m in the foothills of the Alps to roughly 10 m in the area around the airport.

The 10 m thick layer of Quaternary gravel sits on a Tertiary dividing layer of low permeability which separates the Quaternary and Tertiary layers and their groundwater systems.

The separation into two groundwater aquifers is testified by both the difference in the chemistry and the age of the Quaternary and the Tertiary groundwater, and the slightly higher artesian water level of the Tertiary water.

Thanks to these geological characteristics in the airport's surrounding region, groundwater in tertiary deposits at deeper levels is protected from possible pollution in groundwater nearer to the surface.

The groundwater from the Tertiary aquifer is used to supply local communities with drinking water. When work on the foundations of the buildings was carried out, special care was taken to ensure that the thick dividing layer was not damaged and that this groundwater would remain protected once the airport began to operate.

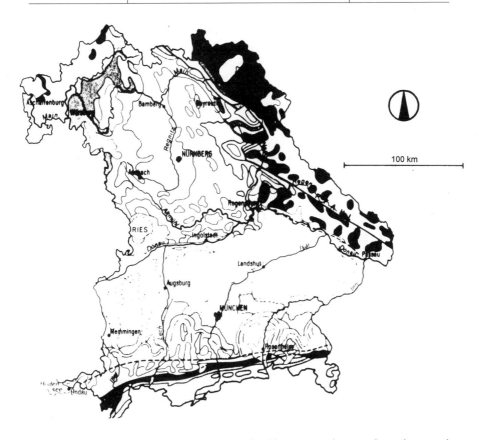

Fig. 2. Geological survey of Bavaria (published by Bayerisches Geologisches Land-esamt, Munich)

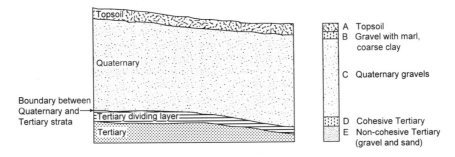

Fig. 3. Hydrogeological section

Groundwater in the Quaternary deposits flows from the south-west to the north-east through the airport site. It has a fall of 2% and flows at a rate of between 2–5 m/day. The average permeability k of the Quaternary gravel is approximately 10^{-3} m/s.

The natural level of groundwater in the Quarternary deposits is high and, because the gradient of the Tertiary substrata is flatter than the gradient at ground level, is nearer the surface to the north. The groundwater level is subject to natural fluctuations of between 1–1.5 m, with the result that it rises periodically to ground level in the northern part of the airport.

INTERVENTION IN THE WATER REGIME

Regulation of the groundwater

To ensure the usability and the safety of the runways, taxiways and aprons, there has to be an adequate distance between the water table and the subgrade on which the paved areas are built. The surface of the groundwater must not rise above the frost penetration depth, which at Munich Airport was taken to be 1 m.

The groundwater level had to be lowered. This was achieved by installing systems of drainage channels in the north and south which reduced the natural groundwater level by roughly 1.5 m within the airport perimeter. At the same time, the effects on the groundwater regime outside the airport were limited as far as possible.

The permanent draining of groundwater would cause an extensive drop in the groundwater level to the north of the airport. To avoid this, the groundwater which is drained in the drainage channels to the north of the airport, is fed back into the ground through an infiltration plant so that the groundwater returns to its original level. The infiltration plant consists of a pumping station with a capacity of 120–450 l/s, a 6 km long distribution line, 44 control shafts, and 132 infiltration wells, each 10 m deep. All of the drained groundwater which flows into the pumping station via the water diversion running from south to north is fed back evenly into the Quaternary aquifer.

Re-routing of watercourses

Erdinger Moos is an area full of small streams which served as drainage channels while the area was cultivated and kept the groundwater at a level favourable to agriculture. These streams crossed the site chosen

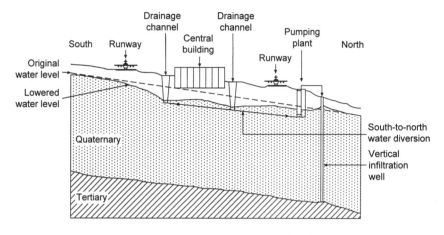

Fig. 4. Section showing regulation of the groundwater level

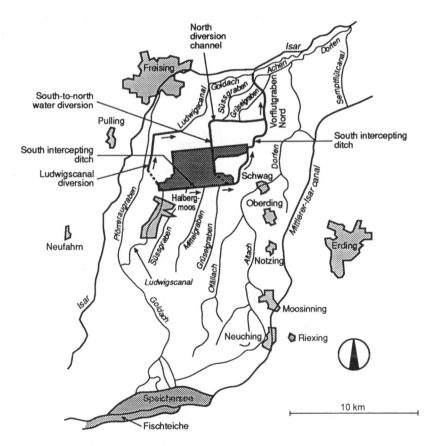

Fig. 5. Plan showing re-routing of watercourses

for the airport, which meant that they had to be diverted and re-channelled.

The receiving water for all surface and underground water is the River Isar. The mean discharge volume of the streams is less than $0.5 \text{ m}^3/\text{s}$; many have a discharge volume of less than $0.1 \text{ m}^3/\text{s}$.

The Ludwigskanal follows a route around the airport in the west. The remaining bodies of water are fed into the newly built south intercepting ditch along the airport's southern perimeter, are channelled via the east intercepting ditch around the airport to the east, and then on into the River Isar, via the new north drainage ditch.

During the planning and construction of the new water channels, one focus was on designing water management measures that were as natural as possible under the prevailing circumstances.

Drainage of waste water and precipitation

The construction of the airport led to the creation of roughly 500 hectares of paved areas that affect drainage in a region with

☐ small bodies of water with a low loading capacity
☐ a highly permeable subgrade, and
☐ groundwater that is only a short distance below ground level.

The waste water generated at the airport can be divided into the following categories:

☐ domestic sewage, for example sewage from the passenger terminal, offices, personnel facilities, canteens and restaurants, kitchens, and the catering companies
☐ industrial sewage produced, for example, in the maintenance hangars, where the washing of aircraft produces considerable quantities of waste water
☐ combined water consisting of a mix of waste water and precipitation
☐ rainwater from roads and parking areas
☐ unpolluted water from runways, taxiways and aprons
☐ rainwater containing light liquids that runs off aprons on which aircraft are parked
☐ water containing de-icing fluid following the de-icing of paved areas and aircraft during winter operations.

The different types of waste water are collected and drained off in a system of sewage, combined water and rainwater drains.

The rainwater overflow tank is only allowed to discharge once in a period of five years, and the waste water that drains into the sewage and combined water sewerage system is treated at the sewage treatment plant operated by the Erdinger Moos municipal sewage association.

The composition of the waste water is much the same as normal municipal waste water. The sewage from kitchens, canteens, restaurants and catering companies passes through grease traps.

Sewage from aircraft is fed into the sewerage system at a special unloading point. The volume is small, so this kind of handling is acceptable.

The sewage produced by the washing of aircraft in the maintenance hangars is pre-treated using a combined chemical and physical demulsification and flotation process to ensure that it complies with legally defined limits before it is fed into the sewerage system.

Drainage areas on the aprons where aircraft are refuelled and turned around and where rainwater bearing kerosene and oil can accumulate are connected to the combined water-sewerage system via light-liquid traps that incorporate a coalescence stage. Rainwater discharge in excess of a defined critical volume is channelled into the rainwater sewerage system via rainwater overflows to the storm water settling tanks, which are equipped to retain any residual light liquids. The overflow from the storm water settling tanks then flows via the south-to-north water diversion, and from there into the network of waterways to the north of the airport.

Car parks and roads in the northern and southern technical areas are connected via storm water settling tanks to the combined water-sewerage system until critical rainfall occurs. This prevents the surge of pollutants typically encountered with the onset of rain from entering the local surface and underground water. Overflow from the settling tanks discharges into the local bodies of water.

The rainwater draining off the runways, taxiways and taxiing areas on the aprons is unpolluted, so it is allowed to drain off the shoulders of the runways and taxiways into the ground, where it partly percolates downwards and partly flows via depressions into the drainage channels. From the canalised taxiing areas on the aprons it drains through rainwater channels and settling tanks into the receiving water. In winter the operating conditions are different and will be discussed separately below.

GROUNDWATER POLLUTION CONTROL

Prevention of groundwater pollution by chemical de-icing agents

Trouble-free airport operations during winter weather conditions require that flight operations areas are kept sufficiently clear of snow and ice and provide adequate traction for safe take-offs, landings and taxiing.

In addition to using snow ploughs, blowers and sweepers for mechanical clearance and the spreading of sand to improve grip, the airport uses chemical de-icers based on glycol and urea to de-ice compacted snow tracks and layers of ice. One of the requirements attached to planning approval for the airport was that the methods used to de-ice flight operations areas were not to pollute either surface water or groundwater at the airport, or to overtax the sewage plant at Eitting. This resulted in the implementation of a combined multiple-stage scheme comprising operational and constructional measures.

On the operational side, the use of chemicals is kept to the necessary minimum. This is achieved by intensifying mechanical clearance and using improved winter services equipment; by increasing the use of sand; by improving the information available on the condition of operations areas through the use of a glazed frost early warning system; by drawing on local and regional meteorological data and incorporating it into the winter services strategy; by specially training winter services personnel; and by using selected chemical de-icers sparingly and with greater precision through the classification of operations areas according to priorities under consideration of the different operational and safety requirements. One option also examined was to heat the runways to de-ice them, but this project was abandoned for economic reasons and because a number of unclarified technical questions made it potentially risky.

The 'in-situ decomposition system' (IDS) installed along the taxiways

De-icing agents deployed by winter services flow along with meltwater and rainwater over the edge of the taxiways, percolate through the topsoil and enter the groundwater, which is roughly 2 m below ground level. The only way to prevent pollution of the groundwater is to break down the de-icer into end products that do not endanger the water quality as they infiltrate the ground.

This idea led to the development of the in-situ decomposition system, or IDS. The action of existing soil bacteria was to be supported by optimising the soil structure in the percolation zone between ground level and the surface of the groundwater so that glycol-based de-icers would degrade to produce water and carbon dioxide.

To achieve this degradation the bacteria required a certain type of soil structure and sufficient time. However, the type of soil structure normally encountered at Munich Airport is such that the time it takes for water to drain through the percolation zone at the edges of the taxiways is not enough to prevent pollution of the groundwater through de-icing agents.

Field and laboratory tests led to the development of a special structure to be installed in lateral strips along the taxiways. The path taken by meltwater containing de-icer as it percolates into the ground was extended substantially by building an obstructive layer that diverted the direction of flow from the vertical to the horizontal in the percolation zone. In addition, the time taken by the meltwater was increased and its route extended and evened out through the installation of alternating banks of highly permeable gravel and less permeable sand.

Tests in the field and in labs indicated that it was possible to achieve a degradation rate of 97% when propylene glycol-based liquid de-icers were used. The efficiency of the IDS is monitored by testing the quality of the groundwater at monitoring points within the airport grounds.

When the actual construction work was carried out, the contractor proposed using a special geotextile sealing mat to enhance the system. The mat consists of two pinned pieces of bonded fabric between which a 3 mm layer of bentonite powder is sandwiched. Once the mat is in place, it absorbs water and swells the activated calcium bentonite so that it forms a 10 mm thick sealing layer with a sealing value of $K = 1.10^{10}$ m/s. A total of 660,000 m^2 of sealing mat were laid at a rate of 7000 m^2 per day.

Measures to dispose of waste water containing de-icer in the vicinity of runways

Urea-based de-icers are also used on runways and high-speed taxiways, but these are not adequately degraded by the IDS. Originally, the

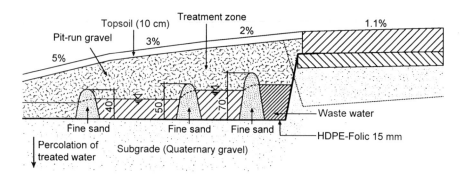

Fig. 6. Section showing the IDS

use of urea on runways was regarded as unavoidable because of the frequency of freezing rain at Munich Airport and the high degree of friction required on runways. Meltwater containing urea therefore has to be collected, drained off and treated.

The shock load from glycol and urea contained in rainwater and meltwater running off these areas is such that the water cannot be treated together with other airport waste water immediately at the municipal sewage association's plant at Eitting. The waste water produced by de-icing operations therefore has to be intercepted, drained off and stored temporarily.

To achieve this, catch-water channels were fitted along the length of the runways, and a separate sewerage system 20 km long in all was installed. This channels the trapped de-icer-laden water into an 80,000 m^3 retention facility, with a 60,000 m^3 underground reinforced-concrete tank, and a 20,000 m^3 surface basin created with a geotextile sealing mat. From here the water is discharged evenly in terms of pollutant concentration and flow rate into the Eitting sewage plant for processing with the local community's sewage.

Prevention of groundwater pollution by hazardous substances

The aviation fuelling system

The permeable subgrade and the high groundwater level meant that special attention was devoted to groundwater pollution control in the construction of all parts of the airport at which substances hazardous to the water quality are used. The most extensive of these was the fuel farm, at which 500,000 tonnes of aviation fuel are handled per year. In cooperation with the regional water board and a technical expert, a scheme based on redundancy and constant monitoring was implemented to achieve the best possible degree of protection.

The fuelling system comprises three main components: storage, transfer to storage, and transfer from storage.

The airport has four storage tanks, each with a capacity of 4500 m^3 and constructed as a 'tank within a tank' with an outer jacket. If the inner tank leaks, the whole of its contents are trapped in the outer tank. A vacuum is used to monitor the double bottom of the tank to ensure that it is still sealed. During filling, two independent overfilling indicators exclude the possibility of overfilling the tanks. The tanks are filled from tanker trains, a long-distance pipeline, and road tankers.

All the facilities involved in transferring kerosene to the storage tanks, such as the rail link, the tank car pumping station, the tank truck pumping station, and the distribution station (manifold), are built in concrete tanking or sealed containers. The concrete tanking is equipped with oil sensors to detect any kerosene leakage from the fittings.

Areas in which waste water polluted with kerosene can occur are connected to a separate sewerage system which feeds a rainwater retention basin. The waste water passes through downstream oil traps and is discharged in controlled quantities.

The piping system at the fuel farm is located above ground, allowing any damage to the pipes to be detected by means of visual inspection.

The fuel is transferred out of the tanks via the hydrant pumping station, likewise embedded in tanking as described above. Behind the pumping station is an underground system of parallel pipes which carry the fuel to the fuelling points, so-called hydrant pits. The parallel pipes ensure that a supply is still available if maintenance work is necessary or a failure occurs. A galvanic protection system makes it possible to detect the slightest damage to the insulation. The large areas of concrete paving covering the pipes on the aprons also provide additional protection.

All the fittings of the supply system are located in concrete shafts whose floors and walls are covered with a kerosene-proof coating.

In addition, the whole system is checked for leaks when the airport is not in operation, in other words at night. Testing is carried out using a pressure fluctuation method, which essentially involves the measurement of the pressure change twice in succession, in a defined and sealed-off section of piping, over a specific period of time. Differences in the pressure curves plotted are evaluated to detect leaks.

The final link in the fuelling chain are the hydrant pits, of which there are 174 on the west apron and 24 on the cargo apron. Here, too, special emphasis was placed on the control of water pollution.

The pit casings were specially developed for Munich Airport in close cooperation with the regional water conservation office and the Bavarian office of the TÜV. Sometimes, small amounts of kerosene are spilt during fuelling operations, and a complex and monitorable sealing system precludes the possibility of a leak from the hydrant casing into the ground.

The whole of the aircraft fuelling system is controlled and monitored by complex process control systems. In addition to measurement data, alarms are sent to the fuel farm monitoring centre and are indicated visually and acoustically to ensure that the duty personnel, who are present around the clock, can respond immediately by taking appropriate action.

Other installations handling hazardous substances

All other airport installations that handle substances potentially hazardous to water, e.g. petrol stations, the chemicals store in the supply centre, the de-icer store, and the vehicle washing areas, were designed and built to offer the best possible protection for groundwater by providing redundant collection facilities and automatic monitoring and collection systems.

To keep dangerous substances such as chlorinated hydrocarbons out of the airport and away from the waste water system, the airport agreed with the users of the aircraft maintenance facilities that these substances should not be used.

Prevention of groundwater pollution through sealed surfaces and pipes

To prevent pollution of groundwater through the use of hazardous substances, paved surfaces and piping have to be permanently sealed. A number of examples describing how this was achieved at Munich Airport are provided below.

A. Passenger handling, cargo and maintenance aprons and de-icing points   These areas are paved with a 36–40 cm layer of concrete on top of a 20 cm layer of lean-mix concrete. The expansion and contraction joints are sealed with a kerosene-resistant filling compound or sealing strips. The joints are checked visually for signs of changes, damage or leakage during regular inspections of the paved areas. Checks are carried out on the behaviour of joints as part of a measurement and research program underway at Munich Airport into the movement of joints, which aims to advance joint-sealing techniques. Hollow neoprene strips, 10 mm wide (Phönix, Metzler), are pressed into the 8 mm cut contraction joints.

We expect the seals to remain adequately tight for 10–15 years. Intersections constitute a special problem, and here the seals have to be cut, bonded and inserted with care.

Correspondingly large seals are used in the 25 mm wide expansion joints, and what is important here is adequate torsional rigidity. At present, seal manufacturers are looking into ways to improve this by bonding wide seals of this kind to the edges of the joints.

B. Runways and taxiways   These areas are paved with a 40 cm layer of concrete on top of a 15 cm layer of lean-mix concrete.

The joints are sealed with a bitumen-based hot joint-sealing compound. This compound has to be resistant to kerosene, and must be able to expand while remaining bonded with the edges of the joint. The joints have a life-span of five years, after which they have to be completely renewed.

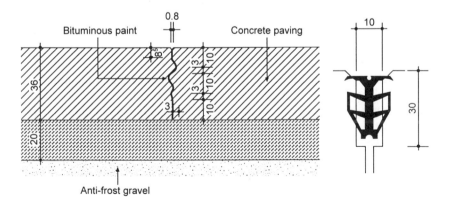

Fig. 7. Jointforms and the cross-sections of sealing strips

Since its was unclear whether sealing strips would adequately withstand the suction pressure that arises when aircraft taxi over them at high speed, this tried-and-tested method of joint sealing was used on the runways and taxiways.

C. Fuel farm, fuelling points, fuel transfer areas In these areas, surfaces on which fuel spillage may occur are either paved with impermeable concrete and the joints are sealed with a sealing compound or sealing strips, or they are covered with concrete sett paving and the joints sealed with fuel-resistant seals applied using a technique approved by the water authorities.

D. Underground pipelines carrying waste water containing substances hazardous to water quality Sewers must always be free of leaks, and this is tested using a 5 m column of water. The sewers into which rainwater carrying fuel can drain are additionally sealed. The insides of the pipes are painted with a fuel-resistant coating. The sewers in the vicinity of the central de-icing areas are pressure pipes which are tested for leaks with a 50 m column of water. The sewers and drains are subject to constant inspection and are checked for leaks every five years. Table 1 provides an indication of the quantities of waste water, fuel and chemicals, and gives an idea of the pollution potential to be combated by groundwater protection measures.

Controlling groundwater pollution during construction work

Protection of the Tertiary dividing layer

In the construction of underground structures, for example S-Bahn railway tunnels, utility tunnels, basements, tanking for underpasses, etc., care had to be taken during building pit closure to guard against foundation water pressure from the Tertiary floor, but without lowering the Tertiary water pressure. This was achieved by installing an underwater concrete floor or a gravel surcharge. To avoid changes to the groundwater regime through dewatering and to limit the range of any changes arising in connection with large, long-term building pits (for example, in the case of the terminal building and the maintenance hangars), narrow-wall closures were used which extended into the Tertiary layer and form a water-tight trough.

If sheet pile walls were vibrated into the Tertiary dividing layer, the channel produced by dragging was filled up with clay suspension via jetting lances mounted on the sheet piles.

Table 1. Quantities of waste water fuel and chemicals from airport operations

Waste water	
Domestic sewage (from sanitary facilities, kitchens)	400,000 m^3 p.a.
Industrial sewage (from workshops, aircraft maintenance facilities, washing facilities)	100,000 m^3 p.a.
Polluted precipitation running off paved areas (from de-icing, fuelling, traffic)	500,000 m^3 p.a.
Fuel	
Fuel supply system (fuel farm, hydrant system)	18,000 m^3 in storage 500,000 m^3 p.a. turnover
Supply centre	2000 m^3 in storage
Petrol stations (7 in all)	1000 m^3 in storage 50,000 m^3 p.a. turnover
Chemicals	
Supply centre	100 m^3 in storage
Aircraft maintenance facilities, workshops, temporary store for hazardous waste	100 m^3 in storage
De-icing agent store	1000 m^3 in storage 4000 m^3 p.a. turnover

Protection of the groundwater regime

The water pumped out during the drainage of closed and open building pits (which at times required a pumping capacity of 8–10 m^3/s) was returned to the groundwater within the construction site via percolation ditches.

Protection against substances hazardous to water quality

The monitoring of all construction areas to ensure that fuel was stored and transferred to the construction site vehicles correctly and that any fuel spillage occurring as a result of accidents was cleared, were constant tasks of a co-ordination office. Central vehicle washing facilities which incorporated sludge and oil traps ensured that no uncontrolled vehicle cleaning on the extensive construction site could endanger the groundwater.

One engineer from the co-ordination office, a second engineer from the airport's technical department, and an official from the water authority were responsible for these tasks on the construction site.

Through the co-ordination of drainage for all individual construction locations, and through monitoring and the preservation of evidence it was ensured that the official requirements were met. The water authority was also involved in this process.

MONITORING THE WATER

Monitoring the quantity of groundwater

To monitor and gather evidence on the state of the groundwater and assess the influence of the groundwater management measures undertaken by the airport, an extensive network of monitoring points – 250 in all – has been checked and evaluated since 1974. There are 57 standard monitoring points, plus 94 which are checked in connection with specific events, and a further 99 which are checked in the case of exceptional events. The water authority publishes a yearly report detailing the results. Before the airport opened in 1992, a comprehensive analysis was carried out and the results were presented as a contour map of groundwater levels. The results proved that the effects lay within the scope of the planning permission requirements.

Monitoring of groundwater quality

Fifteen groundwater monitoring points are checked on a quarterly basis: the main test is carried out in April, and three short tests are conducted in July, October and January.

The parameters analysed are designed to detect possible pollution by fuel and chemical de-icers.

The following parameters are analysed.

☐ In situ tests (M = main test; S = short test)
– Colouration (M, S), cloudiness (M, S), odour (M, S), temperature (M, S), pH value (M, S), conductibility (M, S), oxygen content (M, S)
☐ Laboratory tests
– Colouration (M, S), cloudiness (M, S), odour (M, S), temperature (M, S), pH value (M, S), conductibility at 20° (M, S), oxidability with KMnO (M, S), TOC (M, S), BOD_5 (if TOC >3m g/l) (M, S), polycyclic aromatic hydrocarbons (in accordance with the drinking water laws) (M), surfactants (anionic) (M), spectral absorption coefficient at 254 nm and 436 nm (M), acid capacity (up to pH 4.3) (M, S), alkaline capacity (up to pH

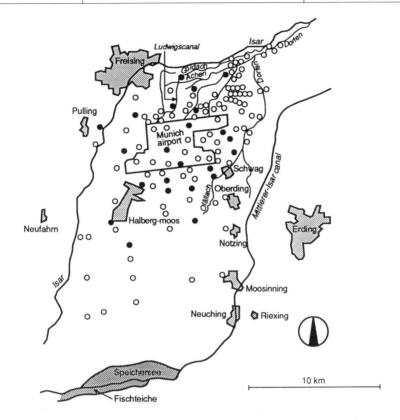

Fig. 8. The network of monitoring points used to check the groundwater level

8.2) (M), calcium (M, S), magnesium (M, S), sodium (M), potassium (M), ammonia (M), total nitrogen (M, S), nitrite (M, S), chloride (M, S), sulphate (M, S), phosphate (M), silicic acid (M), iron (M), manganese (M), zinc (M), cadmium (M), boron (M), short biological test (e.g. luminous bacteria or daphnia test) (M).

In addition, gas chromatography tests are carried out:

☐ Detection of hydrocarbons etc., in the pentane extract using an FID detector or MS (M, S).

☐ Detection of volatile halogenated hydrocarbons, including R 113 (headspace analysis) using an ECD detector (M, S).

☐ Detection of nonvolatile halogenated hydrocarbons up to and beyond Chlophen A 60 from the hexane or pentane extract (approx. 1:200) using an ECD gas-chromatogram (screening) (M).

ANNUAL OVERVIEW 1991 – 1993						
Test Number	Units	Method	P3112/1	P3112/2	P3112/3	P3112/4
Date of Test			12.08.91	29.06.92	13.10.92	19.01.93
Measuring Point			P3112	P3112	P3112	P3112
FIELD SAMPLING						
Colour		11	Grey	Colourless	Colourless	Colourless
Turbidity		21	Turbid	Clear	Clear	Clear
Odour		31	Odourless	Odourless	Musty	Musty
Temperature	°C	60	12.2	10.8	12.0	9.6
pH		80	6.97	7.02	6.92	7.26
Conductivity at 25°C	µS/cm	71	744	887	994	829
Dissolved Oxygen		100	Fixed	Fixed	Fixed	Fixed
LAB SAMPLING						
Colour		10	Yellow	Colourless	Colourless	Yellow
Turbidity		20	Clear	Clear	Clear	Clear
Odour		30	Odourless	Odourless	Odourless	Odourless
pH		81	6.83	6.75	6.86	6.98
Conductivity at 25°C	µS/cm	73	824	944	966	801
Oxygen Content	mg/l	101	0.6	0.4	0.5	0.4
Coefficient of Adsorption 436 nm	m–l	110	1.4	1.0	–	–
Coefficient of Adsorption 254 nm	m–l	110	24	24	–	–
Surfactants	mg/l	1090	<0.05	<0.05	–	–
Permanganate Oxidation Value $KMnO_4$	mg/l	980	27	25	26	27
DOC	mg/l	1010	7.6	6.9	12	7.8
BSB5 (O_2)	mg/l	1020	2	<1	<1	<1
Ammonium	mg/l	320	0.80	0.68	–	–
Nitrite	mg/l	310	0.06	<0.02	0.02	<0.02
Nitrate	mg/l	290	15	21	32	22
Total N	mg/l	280	8.0	7.0	10	8.0
Kjeldahl–N	mg/l	281	5.0	2.4	2.5	3.3
Chloride	mg/l	170	25	32	31	24
Orthophosphate	mg/l	340	0.05	<0.05	–	–
Sulphate	mg/l	220	64	81	93	77
Acidity to pH 8.2	mmol/l	400	1.91	1.80	–	–
Acidity to pH 4.3	mmol/l	370	7.4	7.3	7.6	6.1
Sodium	mg/l	450	6.5	13	–	–
Potassium	mg/l	470	12	5.1	–	–
Magnesium	mg/l	530	22	24	25	23
Calcium	mg/l	500	130	160	160	120

Fig. 9. Extracts from the groundwater monitoring log

Monitoring water quality in watercourses

A biological examination of the water in the streams in the area is carried out once a year to provide evidence of the quality. Samples are taken three times a year at 15 quality monitoring points in order to check the following chemical and physical parameters: appearance, odour, water temperature, conductivity, TOC, BOD, hydrocarbons in solution, absorbed in seston, total ammonia, chloride, nitrate, phosphate, pH value, oxygen.

Sampling and analysis are carried out by an authorised laboratory. The results are presented to the water authority for examination and evaluation. To date, there have been no signs that the airport has affected the quality of the groundwater or surface water.

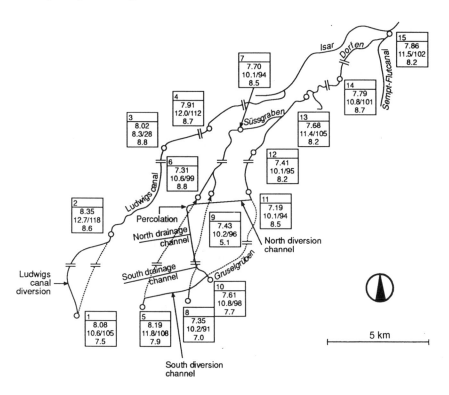

Fig. 10. Locations of the monitoring points used to check the watercourse quality

SAMPLING POINT	1/52	2/52	3/52	4/52	5/52	6/52	7/52	8/52	9/52	10/52	11/52	12/52	13/52	14/52	15/52
Appearance	Clear C/less	Clear C/less	Clear C/less	Clear C/less	Clear Yellowy	Clear Yellowy	Clear C/less	Clear Yellowy	Clear Yellowy	Clear Yellowy	Clear Yellowy	Clear Yellowy	Clear Yellowy	Clear Yellowy	Clear Yellowy
Odour	Negligible							Negligible							
Water Temperature	7.5	8.6	8.8	8.7	7.9	8.8	8.5	7	9.1	7.7	8.5	8.2	8.2	8.7	8.7
Conductivity at 25°C	54	56	59	60	60	82	79	77	83	93	83	82	80	80	78
TOC (mg/l)	3.2	3.5	3.5	3.5	6.6	5.7	5.3	10	6.2	13	6.4	7.3	6.5	6	6.4
BSB5 (mg/l)	1.1	1.2	1	1.1	1.6	1.5	0.9	0.7	1.2	0.8	1	1	0.8	0.8	1
Dissolved Hydrocarbon (mg/l)	<0.05	<0.05	<0.05	<0.05	<0.05	<0.05	<0.05	<0.05	<0.05	<0.05	<0.05	<0.05	<0.05	<0.05	<0.05
Absorbed Hydrocarbon (mg/l)	<0.05	<0.05	<0.05	<0.05	<0.05	<0.05	<0.05	<0.05	<0.05	<0.05	<0.05	<0.05	<0.05	<0.05	<0.05
Total Hydrocarbon (mg/l)	<0.05	<0.05	<0.05	<0.05	<0.05	<0.05	<0.05	<0.05	<0.05	<0.05	<0.05	<0.05	<0.05	<0.05	<0.05
Ammonium (MH_4, mg/l)	0.06	<0.02	<0.02	<0.02	<0.02	<0.02	<0.02	<0.02	<0.02	<0.02	<0.02	<0.02	<0.02	<0.02	<0.02
Chloride (Cl, mg/l)	24	26	27	27	22	24	25	21	26	27	24	24	25	27	27
Nitrate (NO_3, mg/l)	23	23	22	21	18	24	21	36	24	41	16	17	20	18	18
Orthophosphate (PO_4, mg/l)	0.55	0.53	0.45	0.39	0.2	0.05	0.05	0.05	<0.05	<0.05	<0.05	<0.05	<0.05	<0.05	<0.05
pH	8.08	8.35	8.07	7.91	8.19	7.31	7.7	7.35	7.43	7.61	7.19	7.41	7.68	7.79	7.89
Dissolved Oxygen (mg/l)	11.6	12.7	8.3	12	11.8	10.6	10.1	10.2	10.2	10.8	10.1	10.3	11.4	10.8	11.5
Saturate	105	118	78	112	108	99	94	91	96	98	94	95	105	101	107

Fig. 11. Extract from the watercourse monitoring log

Central process control, measuring and control systems used in the water management facilities at Munich Airport

Due to the close ties between the watercourse re-routing, groundwater control and drainage measures, which need to interact smoothly, and the airport's requirements in terms of safe and efficient operation, a water management measuring and control system was installed, which features intelligent substations that use stored program control technology.

All the key discharges are measured and registered, are automatically limited and distributed by the relevant technical equipment, and any failures are detected and reported. All this is taken care of through automatic in situ control equipment and by the water management control centre located in the infiltration plant pumping station.

The control centre can be monitored and controlled remotely via Telekom's telephone system using an acoustic coupler and the private phone connection of the stand-by team. All stations are equipped for the connection of a portable data terminal which allows on-site access to the data and, thus, intervention in links and programs.

The water management system is one subsystem in the central process control system for the whole of the airport's installations. This means that the water management computer in the infiltration plant pumping station is a sub-centre.

The measurement and control systems and the central process control system not only ensure that the requirements tied to planning approval are fulfilled, they also help to optimise the operational phase: a status report can be obtained at all times and appropriate action taken.

In addition, malfunctions are reported automatically and their nature and location are specified, allowing immediate and precise intervention.

A mass of measurements of water levels and flow rates allow the computer in the infiltration plant pumping station to determine the discharge volumes in connection with the watercourse re-routing, groundwater control, drainage and water supply measures. The data collected is evaluated; the IT systems then issue control commands which trigger discharges of appropriate volumes into the north drainage ditch, the Süss, Mittel and Grüsel ditches in the north, and the draining of water into the west and east percolation pipe runs.

In the event of an oil spillage, an appropriate alarm is triggered which closes the gate automatically in the relevant area and keeps it closed until the spill has been cleaned up.

The waste water discharge into the main intercepting sewer for the sewage plant is monitored at a measuring station at the crossover point into the drain of the municipal sewage association, and its pH value, its conductivity and its TOC are analysed chemically.

The drinking water consumption at Munich Airport is measured in the west and east shafts in the south of the airport and the data collected is processed in the computer. Pressure manometers ensure that an interruption to the water supply from the Moosrain municipal association causes the valves of the emergency line to Freising Süd to open, and water can be drawn from here, although requirements must then be reduced.

Organisation and costs of monitoring the water

The planning, construction, running and operation of the water facilities are the responsibility of the Civil Engineering and Water Management department. This department is part of the airport's Technical Division and belongs to the principal department Planning and Construction.

The head of the department fulfils the role of water pollution control representative as required by section 21 of the German Water Regime Law. The duties of this role are

☐ to ensure that water pollution control requirements are complied with
☐ to secure appropriate treatment of waste water
☐ to report to corporate management once a year on water pollution control measures that have been implemented or planned.

The planning, construction, running and operation are managed by sections responsible for surface and underground water (two graduate engineers, three technicians and four labourers), and sewage and waste water treatment (two graduate engineers, three technicians and two labourers).

The implementation of the relevant tasks is contracted out to outside firms and laboratories.

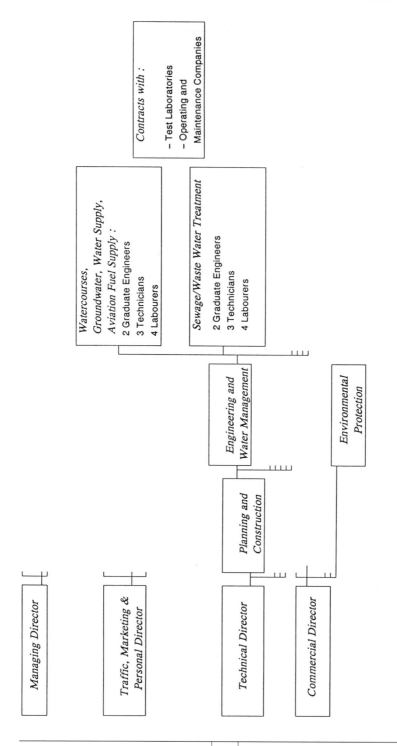

Fig. 12. Organisational chart for water pollution control

Regular reports are sent to the airport management and the authorities, describing the current state of development, the measures to secure evidence of water quality, and any noteworthy occurrences.

Approximately DM350 million have been invested in water management and water pollution control measures. The running and operating costs amount to roughly DM6 million per year.

CONCLUSION

The new Munich Airport and its requirements in terms of water management and control constituted an intrusion into a rural area, an area known as Erdinger Moos, and a water regime typical in terms of its situation and structure for such an area.

At first one might think that what we have here are two irreconcilable opposites, but they have successfully been united through an integrated water management scheme based on the redirection of watercourses, the control of groundwater, the implementation of drainage and pollution control measures, and the installation of water supply facilities.

In the areas of surface and groundwater pollution control in particular, innovative solutions were designed and put into practice. The overall result makes it fair to say that all those persons involved, be it from the airport operating company, the planning offices, or the water authorities, have succeeded during the planning, the legal proceedings, the construction work and the operation of the airport in keeping the impact on the water regime at the airport as minimal as possible.

Waste management and airports

PETER JONES, Director – Development and External Relations, Biffa Waste Services Ltd, Coronation Road, Cressex, High Wycombe, Buckinghamshire, HP12 3TZ, UK

INTRODUCTION

'Waste is everything you make which your customers don't buy'.

This paper takes you through the issues of waste management as we see them (principally here in the UK) in terms of the structural issues and the way in which discussions are being implemented specifically in the UK framework directives as they are appearing from Europe. The process issues are about:

☐ the nuts and bolts activities that we perceive to be underway or planned from a logistics point of view
☐ the umbrella of specific legislation which is impacting on airports. It is a carrot and stick process.

The old hierarchy of waste management is illustrated in Fig. 1. Biffa Waste Services is a national operator in waste, with an annual turnover of about £150 million. We are first equal (maybe even the largest) UK integrated company operating on waste. Biffa talks to a lot of major plcs who are now realising that the concept of sustainability and zero impact is a very significant and interesting game. More importantly it's about bottom line performance as well as environmental performance – more and more companies are being driven to become aware of that hierarchy.

WHY BOTHER?

For years since the 1800s and the industrialisation in Western Europe we have been used to measuring industrial activity in monetary terms, balance sheets and profit and loss accounts. Increasingly companies are now realising they also have to measure their businesses in terms of weight. Fig. 2 illustrates some calculations on the costs to this country in weight terms to sustain its 2% or 3% growth pattern, to allow it to operate as a developed society. The UK economy consumes around

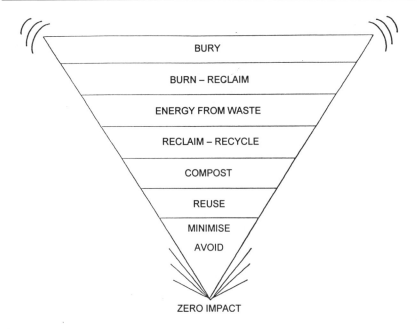

Fig. 1. The waste hierarchy

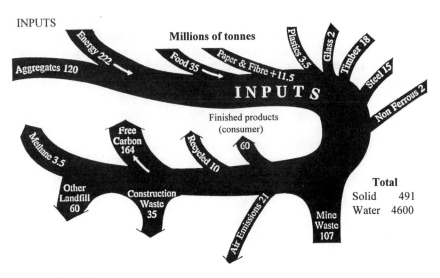

Fig. 2. The UK resource economy

650 million tonnes of input raw resources. This is in the form of millions of tonnes of coal equivalent for energy input processes (such as gas, fuel or oil) as well as the input of timber, aggregate, coal, other raw materials, ferrous and non-ferrous materials.

The United States is a much higher consumer, but the UK as a nation consumes about 0.5 tonnes of food per head and about 0.5 tonnes (as *private* consumers) of everything else each year. So there is an input of 660 million tonnes for 60 million tonnes of private consumer output. That ratio of 10 or 11:1 input of materials to output, is a figure which appears in surveys and waste audits for large integrated companies, whether they are in manufacturing, the service sector, distribution, or logistics. Whatever the industry, the unerring fact is, that it takes 10 or 11 tonnes of input to generate 1 tonne of output.

Now of course that cannot be quantified in airports – in the service business (the output is not weighable so to speak); but as a philosophy it is worthwhile taking companies down that route.

COMMERCIAL AND LEGISLATIVE PRESSURES

The waste management industry has been subjected to what started out appearing as a series of piecemeal initiatives from Europe and from the UK Government at national and local level – but a cohesive framework is emerging as shown in Fig. 3. This concept of sustainability means that each generation leaves the planet as it found it and there is no nett environmental burden for future generations. What the Government is doing is basically making resource consumption and disposal more expensive. Before it started on that route it had to define the rules of the game. As a result of the 1990 Environmental Protection Act, the

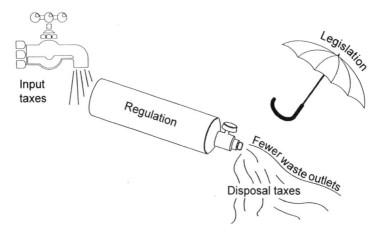

Fig. 3. National waste strategy: a cohesive framework?

UK's Environment Agency was formed in April 1996, which is similar to the USEPA but has a slightly different focus. Control of water, air and solid disposals is to be integrated. That process is going to start with large consumption units such as airports and retail shopping parks. Those sectors are probably going to act with an integrated agency over the next 2 or 3 years. Finally, there are signs of fiscal instruments coming into play in this economy:

☐ input taxes in the form of Producer Responsibility
☐ levies on raw material input
☐ the allocation of responsibility for the retrieval of products to the manufacturers.

Those that manufacture packaging are beginning to be made aware that they could be liable for their subsequent retrieval, if not directly then financially through agencies in the waste sector. Finally the Government is announcing and going to implement, in 1996, taxation on landfill. An integrated strategy is emerging, and the rules of the game are defined for airports in terms of those broad areas. Table 1 shows the water, air and ground emissions and the general umbrella legislation, which forms the tightly regulated environment in which we work.

Table 1. Principal waste directive impacts on airports

☐ Water	EC Groundwater Directive 80/68 Waste Oils 75/439, 87/101 86/278 Sewage Sludges
☐ Air	Select Committee RECS - VOCS ISBN 010293195X
☐ Ground	Litter Regulations 91/157 Batteries EC Hazardous Waste Directive 91/689 Landfill Taxation D Document Hazardous Landfill Directive 1991 Hazwaste Listing 78/319
☐ General	Environment Protection Act 1990 DoE - Duty of Care 1991 Planning Policy Guidance No 23 Producer Responsibility Consultation (expected autumn 95) UN Hazardous Transportation Regulations Framework Directive 75/442/EEC Transfrontier Shipments 259/93 PCBs 76/403

The industry is also regulated by the market place, because bad publicity (as we have recently heard from the oil industry) is now a potent force for market change.

The UK Government is saying that this is about financial instruments – it's about market forces. They have set the legislation and they are not going to put any more significant sums of money into subsidies or other initiatives, other than kick start processes. It is all now down to industry. Table 2 shows a dossier of court cases now building up from the 1990 legislation due to this message being ignored. These are just a few examples, some of which involve well-known companies and relate to incidents which could occur in an airport environment. Airports have grown organically over the last 20 or 30 years, with the result that structures, drainage systems and waste disposal management systems have grown rapidly in an ad hoc way. Environmental auditing and standardisation of procedures which have an impact on the environment is going to become more and more important in the future. If you doubt this, then the courts will reaffirm it for you.

Table 2. Court cases (spring/summer 1994)

Incident	Defendant	Fine
Oil spill	British Rail (Tyne & Wear)	£8,000
Leachate pollution	Devon County Council	£1,550
Gas oil leak	McCarthy Construction	£3,000
Landfill licence infringment	Trailwaste Ltd	£5,000
Ammonia discharge to stream	Christian Salvesen	£16,200
Pig effluent discharge to stream	I Crane (farmer)	£16,200
Clay discharge to brook	Tarmac Construction	£2,000
Injected abbattoir waste into a field near a stream	Cleaning Sevices (Botley)	£4,000

EXTERNALITIES

The Government umbrella waste strategy talks about these sorts of issues in terms of moving up the hierarchy, environmental management statements and recovery initiatives (very similar to what we have seen in Europe, as illustrated in Fig. 4) and there is a concept of internalities and externalities appearing in the UK. The fact is that if you make a product (or generate a service) then you become wholly responsible in the price you charge for that service; not only for the

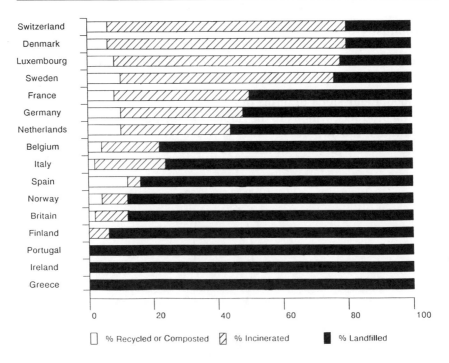

Fig. 4. European waste management

costs 'inside' the business (production, marketing and overheads), but also for the external costs that you impose on society. If one looks at the planning applications for the proposed Heathrow Terminal 5 and the second runway at Manchester Airport, then those externalities begin to be very significant indeed in terms of the potential number of people affected. We are seeing large companies in the UK, such as Manchester Airport plc, develop the concept of an environmental management statement. This is a physical measurement of the impacts they are causing. Typically, many companies are surprised by the potential impacts of their waste and they are realising that they have no comprehensive waste management strategy.

If one looks at the input of resources into airports (or into any other similar economic units) one sees tremendous sophistication in the management of the inbound flow of materials including 'just in time' delivery, intricate warehousing systems and tight control of costs, tonnages and stocks. However, management of the waste stream is typically much more poorly controlled. Management does not know the type and volume of waste being generated, where it is going or

indeed where it could go. Frequently managers do not bother to add up
the cost of it so a cohesive strategy (in terms of environmental management) in airports has to have a three-pronged approach, as illustrated in Fig. 5. The three main factors are:

☐   media emissions in terms of air, groundwater and solids
☐   the products, which are the impacts of the physical materials going
     through the airport
☐   the process (the way in which you select to handle those products).

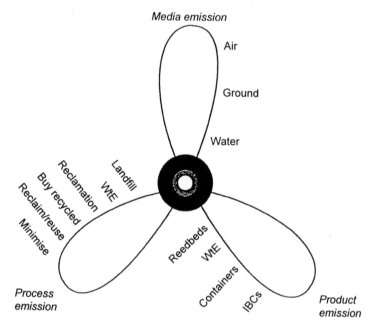

Fig. 5. The zero impact game

All three factors are aspects and ingredients of the sort of impact that is
produced in an airport and the processes could be landfill, waste to
energy, reclamation, recycling or minimisation. One needs to look at
where the waste will be disposed of, which media will be used, what
the waste content is, and finally how the waste is to be treated, handled
and removed from the premises (Fig. 6). The nature of environmental
threats for most commercial enterprises is really a function of the size
of the likely sanction on the one hand e.g. a jail or prison sentence,
balanced against cost on the other. Biffa and other companies within
the waste sector are trying to persuade large organisations and units to
think about where they want to be on this matrix. These companies

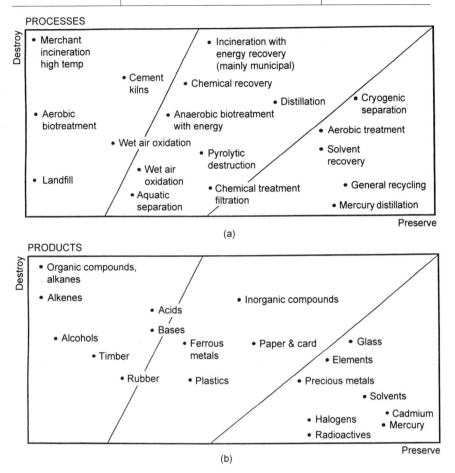

Fig. 6. *Waste options: (a) processes; (b) products*

have to decide whether they want to have a very expensive total zero impact environmental game. The client must define where they are in that matrix in terms of bad publicity, the effect on their customers and the impact on their share price. They have to identify the important parameters (and for many of them this is driven by an inclination to operate at the lowest minimum cost).

LANDFILL

The UK has very cheap landfill. Rates are of the order of £8–£10 per tonne compared to £40+ tax inclusive in Europe, with similar rates in the United States. Approximately 95% of material from most economic activities in this country ends up in landfill but this is changing. The number of landfill sites is decreasing: 5 to 10 years ago there were approximately 4000 active landfill sites, with a fairly broad capability to take everything from toxic, high environmental impact materials right the way through to inert soils. There are now fewer than 500 sites (Fig. 7) (with fewer than 4 significant sites in the vicinity of Manchester airport). The laws of supply and demand have their natural effect in terms of price.

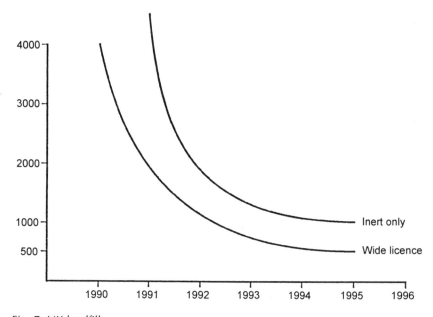

Fig. 7. UK landfills

Throughout the UK the strategy for a particular airport will be a function of how it views that risk. The issues in Gatwick, Manchester, Leeds and Cardiff are slightly different in terms of the economic forces that are powering landfill prices in their particular patch – invariably the key part of the equation in terms of deciding the timing of any changes to strategy.

Another issue is the Government's concern over the slow rate at which the price of landfill disposal is rising and its intention to introduce a landfill tax. Latest information is that having indicated their preference for an ad valorem tax (based on the gate fee paid), it will now be based on the weight of material over a weighbridge. This is going to give the producers of waste an incentive to cut down the volume (Table 3). The level of the tax will probably be of the order of 50–60% of current gate fees. In Manchester, disposal of material from the airport, will probably produce a gate fee of between £12 and £15 per tonne, which is an increase of £8 or £9 a tonne. Gatwick annually generates approximately 20,000 tonnes of waste and Heathrow generates around 30,000 tonnes. So taxation levels of that order mean a third of a million pounds off the bottom line in each major location.

Table 3. Landfill tax

	Typical prices	*Post tax @60%*
Domestic refuse	£6–£15	£9.60–£24
Asbestos	£5–£20	£8–£32
Drummed specials	£20–£50	£32–£80
Inert soils	£Free–£2	£Free–£3.20
Industrial/commercial	£8–£18	£13–£29

WASTE COMPOSITION AND AUDIT

These sorts of messages are urging people to action, as they are a function of the impact tax on inert and construction wastes at a lower tax of £3 per tonne. In the construction industry there are very pro-active initiatives to recycle construction waste (which will possibly avoid tax altogether). Fig. 8 illustrates the composition of the waste

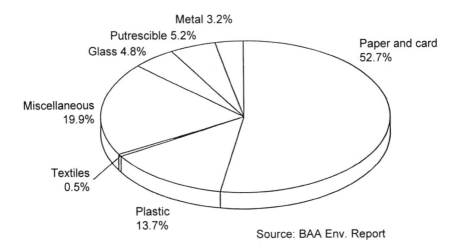

Fig. 8. Gatwick waste composition (20,000 tonnes p.a.)

generated by Gatwick Airport, (which has been taken from their Environmental Report), and clearly shows the predominance of materials such as paper and card, plastic, metals and other recyclable materials. This is significant in that two thirds of this waste is potentially recoverable, and the indications from the environmental report are that approximately 60% of waste generated in an airport environment appears to be coming off the aircraft themselves, as opposed to in-situ generation from the catering facilities. It is estimated in the Heathrow report that as much as 80% was recoverable for end use markets.

The factors to consider include:

☐ where you are discharging your waste to
☐ what it is
☐ how you are handling it.

Secondly, look at the products which are generated. Fig. 9 illustrates this decision making process. There are the two fundamental choices in terms of products. There are those which (preferably) need to be destroyed, and those which should be preserved. Materials such as cadmium, mercury, radioactive waste, and halogens should be preserved and preferably stored for reintroduction and reuse.

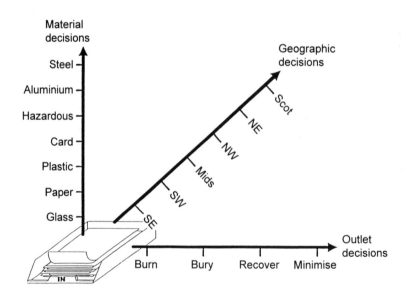

Fig. 9. The waste rubik

The growing awareness of the Rachel Carson syndrome is that it may not matter if we throw away the odd fluorescent lamp as it has a pretty low environmental impact in a landfill site of 30–40 acres. The fact that there are 70 million fluorescent lamps generated in this country each year, means that there are 4 tonnes of mercury going out into the biosphere in some shape or form. This gets into the atmospheric systems, into the rivers, into the grass, and thereby into the food chain. The consequences of this are unknown except that it will (probably) have a long term adverse impact.

It is important that having done your audit, a strategy is identified for these products as well as for inorganic compounds, paper, card and plastics. Whether to destroy or preserve is a function of economics – the move is towards preservation if there is an economic market or there is a processor nearby. The same thing applies to processes. This is interesting in an airport environment, because the scale may be compared to many generators. If the airport is servicing its market place and it is growing, other significant in-situ capital investment may be justifiable, operating on a containment basis without the material being taken outside for processing. The trade-off is between destroying material (which probably means exporting it to merchant high-temperature incineration), or at the other end of the scale there is solvent recovery, treatment filtration, cryogenic separation etc, which can be justified by airport volumes. It may also be possible to operate as the bulk generator for certain waste streams, and bring in merchant material from outside to offset the economic cost of a process (either directly owned by the airport authority or operated by contractors from the waste sector).

RECLAMATION

Recycling is a key issue, which has suffered bad press in the UK, simply because of the volatility of the markets. Many of the solutions in airports are crude, involving compaction devices which occupy minimal space, in 35 yard containers dotted around the complex. To leave it co-mingled for landfill is not a long term solution. That process is beginning to alter. In-situ sortation systems in the airport are starting to appear. Land in airports is very expensive and, in terms of the competing pressures, waste has to take its place in the queue. Increasingly there is a realisation that other systems cannot be added to an old infrastructure. The existing systems must be adapted and a material reclamation facility to process mixed waste within the complex must be considered. Sortation at source is clearly the best solution. The next option where space for an in-situ material reclamation plant is lacking, is to use colour coded containers. Biffa operates a system of collecting waste for landfill during the daytime, and then on a back shift basis using the same truck to collect segregated recoverable waste. This provides raw material for the board or paper plants, with no additional investment in logistics capital, and provides more jobs.

Biffa now recovers 80–90 thousand tonnes of paper annually throughout the UK, which is fed back to the new de-inking plant in Kent. Again this uses standard technology, in conjunction with Svenska Celosa (SCA) – one of the biggest board and paper processors in the world, as illustrated in Fig. 10. SCA are investing in additional reprocessing capacity, which drives demand for these products.

One must consider education a long term process to encourage and explain to staff why sortation is important. Clear colour coding of containers can impact in a positive visual sense in the airport. These schemes can be integrated with a collection round outside the airport so that the truck is also servicing 'cluster' industries such as catering suppliers, engineering companies and others outside the formal 'boundary' to achieve necessary scale. Sometimes the contractor may have to acquire the specialist technology, which is being developed based on American and European experience for glass sortation. Biffa

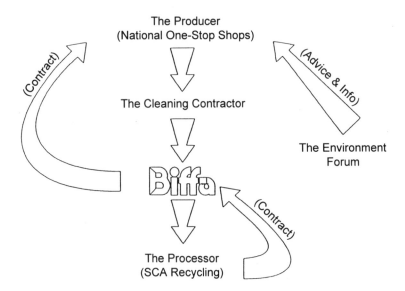

Fig. 10. Paper Tiger – the alliance

is working with major brewers in response to a shift in beer consumption towards glass bottles away from draft beers. We also have an initiative for paper reclamation. Waste contractors need route density and are often prepared to put the containers in (subject to the market conditions) to pull material out at source. The same applies to plastic – the plastics industry is now coming under pressure, under the EC Packaging Directives to pull back more post-consumer, post-industrial and commercial streams.

Many detergents, bleaches, cleaning chemicals containers and other similar substances, which are not contaminated with too hazardous a material are entering airports. These can be shredded by mobile units – this is presently being discussed with a number of trade associations. It is impractical to start these schemes with one collection in Manchester and another one 40 km up the road. Large scale schemes are needed. Airports are big triggers to achieve that scale. The Paper Tiger initiative involves cleaning companies as well – an example of cross-supplier cooperation. The big issue in the whole of the environmental debate is that employees, suppliers, and the management within organisations, rarely have the enthusiasm, verve and approach needed to achieve results until they are forced to do something. Cleaning contractors are an essential part of that process and if they are left out, the result is poor sortation. Where space is not a problem, segregated containers may not be required. The refuse removal in the airport can be wired up in much the same way as you wire up the electrics and the plumbing, with vacuum systems that service from the docking point on the aircraft. Central FLUG of Sweden have installed one of these systems at Arlanda Airport and the waste is therefore out of sight, out of mind, and transported very quickly to central sortation where much firmer disciplines operate over the quality of sortation.

IMPETUS TO END-USE MARKETS

There is no point recovering this material, segregating and recycling it, if all that happens is that it gets thrown into the landfill site. The best foundation is driven by end-use markets as Fig. 11 shows. There has always been a fairly firm market for prices of ferrous and non-ferrous metals, but the markets for paper, plastics and fabric, are now recovering. This is driven by the industrialisation occurring in China and (to a lesser extent) in India, where there are 2 billion people, (6 times the population of the EC) with an economic growth rate of 8% compound over the last three years. There is an engine of growth and hunger for semi-processed raw materials of the recycled nature. They

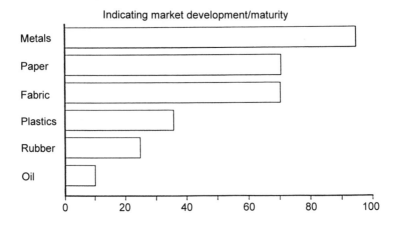

Fig. 11. Kick-starting the markets (examples of recyclable materials)

need these materials because they do not have the energy infra-structure to take virgin materials and process them, so they are raiding available markets for these materials in Europe. There was no market for paper two years ago and it was therefore sold at £5 a tonne. Now that it is in demand, levels have risen to approximately £100–£120 a tonne. Shipping paper to landfill in this country is economically and environmentally wasteful. The waste sector must provide the logistics support to enable cost cuts and minimise environmental impact.

Figures 12 and 13 illustrate another trend developing in the UK and are from the National Recycling Forum. The aims of this are:

☐ to promote the importance of choosing and buying recycled
☐ to demonstrate the existence of markets for materials
☐ to encourage organisations to lead by example
☐ to 'close the loop' between collection, processing and purchasing
☐ to increase understanding of the quality, reliability and availability of recycled content products and materials.

The National Recycling Forum promotes the buying of recycled materials by pointing out the benefits as follows:

☐ stimulates markets for reclaimed materials
☐ reduces waste to landfill
☐ conserves resources
☐ encourages investment in recycling processes
☐ creates purchaser choice
☐ improves environmental performance

□ demonstrates commitment to good practice
□ helps save money.

In the States the 'Buy Recycled Programme' was explained by a speaker to a fairly eclectic collection of people in industry, the waste

□ Opportunities
  – strategic
  – planning
  – operational

□ Overcome obstacles with
  – product information
  – case studies
  – networking

Fig. 12. Overcoming obstacles

□ *If you are not closing the loop by specifying and buying products with recycled content ... you are not recycling!*

- sort recyclables in your organisation

- arrange for their collection by a re-processor

- buy products with recycled content

Fig. 13. Closing the loop

sector, local government and national government. There were discussions on some of these issues – not to lobby for specific courses of action from personal or corporate self interest but to see how reclamation could be made more viable. An important ingredient is that a waste management policy and strategy in an airport is not just about issues of disposal. To benefit from sustainable markets in these materials, and maintain values (which mean that the cost of disposal is lowered) a reclamation programme must be improved and developed, which will survive long term. It is important to focus on procurement and specifiers. The suggestion is not to buy flooring, roof insulation, tables, carpets, and other material made of reclaimed/recycled material unless it satisfies two important criteria: fitness for purpose and cost.

Significant purchasers of huge tonnages of virgin inputs used in buildings, runways, kerb stones and the like, should take note. There is a growing industry manufacturing products from reclaimed materials, at the same price, to the same quality of specification. The more these materials are used, the greater the demand created; the loop will be closed and the longer a sustainable outlet for this sortation will be possible.

'SPECIAL' WASTES

The strategy described applies to the significant tonnage of arisings of fairly low environmental impact materials, but of course airports also generate toxic substances, for example, paints, adhesives, solvented chlorinated materials and fluorescent lamps (with mercury content). The landfill costs for these materials have increased significantly. There is clear evidence in the UK that whilst it is increasingly difficult to obtain planning and licensing consents for landfill, to get planning and licensing consents for high environmental impact products in the special waste category is going to get tougher still. This is a problem if attitudes do not change, but also an opportunity because the technology and the processing capacity for these materials exists. This is a nascent source of employment and economic growth in the UK. It is a shame that Biffa, as a significant purchaser of environmental products (in the form of landfill liners, gas landfill pumping systems and leachate systems) buy most of the required equipment from the Swiss, Germans, Italians, Swedish and Americans. That is because these countries have operated within a much higher regulatory framework for a lot longer. Tight environmental barriers are an opportunity rather than a problem!

The following list of substances are from the EC's List One of Special Wastes, which require special treatment

☐ mercuric compounds
☐ oils/hydrocarbons
☐ halogens
☐ organo tins
☐ pharmaceuticals
☐ insecticides
☐ cadmium
☐ pesticides
☐ lead acid batteries.

All those involved in the waste industry should be familiar with this because the range and breadth of special waste products in the UK will be extended quite dramatically. Some of these substances will become very expensive to dispose of. The manufacturers are now looking at water based solutions and other similar initiatives to lessen the impact. Severn Trent Water Authority have demonstrated the tonnage impact of non-agricultural herbicide applications entering the drains as run-off from gardens (and industry). There are significant surges in April and May on herbicides and pesticides. It has always been a source of irony that the water industry in the UK is spending billions of pounds taking these materials out, when the turnover of these products is measured in tens or hundreds of millions of pounds. Clearly if the manufacturers of these products were forced to carry the cost of removing them at the end of their life cycle, they would not stay in business (Fig. 14). What they ought to be doing is questioning why those products are in the portfolio. Is there a strategy for their handling and their management? The regulator will progressively start cutting back the discharge limits.

ALTERNATIVE STRATEGIES

Airports are being lead towards a philosophy of total containment. In-situ treatment plants, oxygenation plants and reedbed technologies are becoming common. The first reedbed was constructed at Heathrow last summer, and pond systems have been established at Gatwick. Biffa is working on reedbed technologies in conjunction with Severn Trent – principally in relation to mine effluent issues. The bio-organic treatment route is beginning to interest many companies.

There are large hotel and catering chains that operate motorway service areas (which can parallel the sort of catering environment that exists in

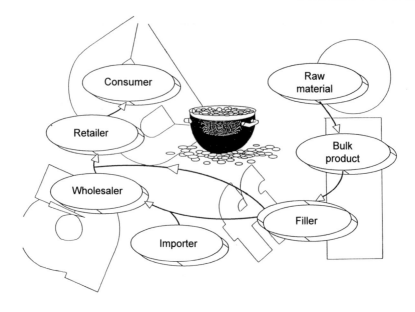

Fig. 14. Where in the chain?

airport terminals) who are looking at vacuum toilets, and white and grey water systems (which is a separate plumbing system to take washings from trucks and low grade cleaning activities for use in toilets). This may be a little difficult in terms of passenger throughput, but is certainly acceptable for staff blocks – especially when the issues are explained. Biogas treatment systems are also in this category, and combined heat and power plants using the methane gas generated from vacuum toilet systems allow reduction of the power burden on a significant airport complex. Gas management systems are now available.

The issues in packaging are returnable packaging for intakes, minimisation initiatives, bulking up, moving from 25 litre plastic containers to intermediate bulk containers, or even tanker deliveries (for things like detergents and bleaches). Not all these necessarily apply in an airport context, but mass balance analysis to understand what is going where and when will help.

CONCLUSION

The waste sector is the grease between the two wheels of supply and demand. This is a sector where all the sophistication on the inbound logistics management (for virgin material inputs) is going to start being applied to what is thrown away at the back end, as illustrated in Fig. 15. The chances of achieving a good end result are diminished where this process is not understood. What has not helped in the UK is the fragmented nature of the waste management sector (see Fig. 16). Biffa is one of the largest companies in a sector worth about £3bn. This is the equivalent of the UK's annual spending on bingo (not including the National Lottery), house plants, or newspapers. Clearly this is a ludicrously low amount. As a result there are not sufficient big, substantially capitalised companies – but that pattern is beginning to change. As it changes what is going to happen? Continuing consolidation will produce higher standards and landfill diversion. More segregation, reclamation and integration between the public and private sectors will also produce viable waste to energy schemes. That is an issue in terms of airports where there are lots of combustible materials.

Domestic reclamation systems for hazardous waste could reduce costs by local integration with airports and other large scale producers in the area. This is a community opportunity to reduce the environmental impact that is imposed on the community. The critical mass exists to solve some of the waste and segregation issues of that surrounding community. A lower cost, and a more progressive opportunity for

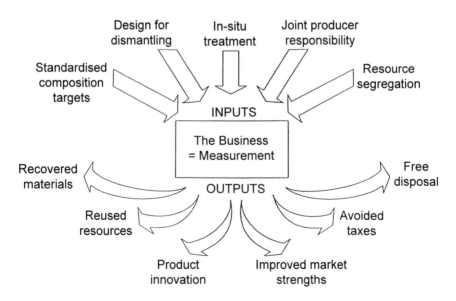

Fig. 15. Processing operations – future

August 1995
sector value £3 billion

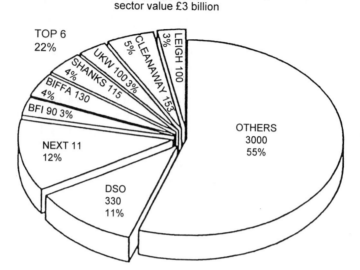

Fig. 16. British waste market: market shares

waste disposal can be provided. The barriers can be broken down as waste does not recognise community boundaries.

This is not some loose 'Save the planet' initiative – it is about competition, credibility, the chances to develop the airport business into the next millennium. Waste strategy is a key element of that approach – it ought to be planned and implemented. Airports mostly focus on sending people around the planet but equally they could be providing clear examples to society – to make sure that it is worth flying around in 2, 3 or 4 hundred years' time – that there is still something worth seeing.

Waste management in the air and on the ground: an ecology model of Vienna International Airport

DR. BRIGITTE MALLE-BADER, Vienna Airport PLC, Postfach 1, A-1300 Wien-Flughafen Vienna, Austria and
NADINE TUNSTALL PEDOE, Scott Wilson CDM, Bayheath House, Rose Hill West, Chesterfield, Derbyshire S40 1JF, UK

Vienna International Airport, faced with increasing mountains of waste and forthcoming national waste management laws, decided to launch a full-scale environmental programme ahead of the implementation of legislation. This programme was initiated as a human-intensive project instead of using costly technology, such as a waste sorting plant. The 'ECO-team', consisting of environmentally specialised public experts, provided a wealth of information for employees, and motivated them through award schemes. The impetus of the programme was carried through to airlines, passengers and associated companies within the airport, such as the fire brigade. The programme has been extremely successful, reducing the amount of residual waste from commercial activities in the airport by 34%, and from the airlines by over 50%.

INTRODUCTION

The management of waste is one of the most important environmental issues presently facing airports. The shear volume of waste produced by an industry the size of an international airport has caused many airports to rethink their waste management strategy. Austria as a country is quite aware of these problems and the population in general prefer to consider the alternatives to facing the increasing demand for sites for landfill and the unpleasantness associated with incineration.

As an undertaking, the Vienna Airport may be compared to a medium-sized town. Housing approximately 200 enterprises, with over 11,000 employees, not to mention 50,000 daily passengers and visitors, its environmental tasks are that of a municipality. The passenger growth at all airports has been substantial. At Vienna, this had risen to 7.8%

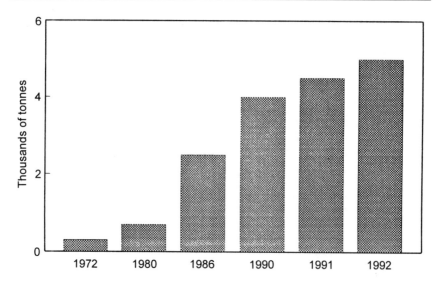

Fig. 1. Waste mountains

per annum by 1992. Fig. 1 shows the resulting growth in the mountains of waste. This waste was identified as consisting of mostly recyclable materials such as metal, paper and glass. In the face of a waste stream literally mushrooming over the years, the airport company 'Flughafen Wien AG' finally decided to enter into a full-scale environmental campaign in 1991.

In keeping with an ecology-oriented waste policy (which differs from ordinary disposal programs), the goal was to put a curb on the waste stream's unbridled growth on one hand, and to become a paragon for other national and international undertakings and entrepreneurial sites on the other.

OBJECTIVES OF THE PROGRAMME

The size of the airport provides both the opportunity and the obligation to practice active environmental protection which goes far beyond air traffic and the associated environmental problems. The fundamental guiding principle of Flughafen Wien AG is to go beyond the legal obligations, to exhaust every possibility, to be sparing with resources.

The proposed 'ECO-model' was not primarily intended to solve the problem through costly investments, such as constructing an expensive sorting plant for the generated wastes, but to place the controlling factor – man – at the centre of all activities, and convince him through information and motivation that he has a duty to care for the environment. It was proposed to raise environmental awareness by applying the 'polluter pays principle', but mostly by relying on supplying adequate information.

STRATEGY OF THE PROGRAMME

The strategy underlying this programme was to involve all of the 11,000 airport employees in the programme through in-depth communication and motivation. 'Flughafen Wien AG' appointed a public-relations expert, specialising in environmental matters, to motivate the airport staff and assist them in the development and enforcement of environment-friendly activities.

The first step in the programme was to hold employee discussion groups, with up to 25 participants in each meeting. These groups were encouraged to develop their own methods of waste minimisation and management. The programme was initiated from the bottom of the staff hierarchy to encourage the staff to develop their self-motivation, rather than following management commands. This led to approaches which were fashioned to specific needs. The airport's amazingly varied infrastructure required that a tailor-made solution was drafted for each area.

As a result, the provided solutions were generally accepted. A determining factor for the programme's success was that all suggestions and ideas which the employees brought forth were carried out by the company-owned disposal enterprise immediately. The prompt implementation of employer proposals are an important aspect of the programme. If large organisations are slow to take up ideas, staff become demoralised and the impetus is lost. The staff members in charge of cleansing and waste disposal were playing a vital role in the ECO-team's public-relations strategy. All waste-related activities (such as providing the necessary facilities for separate collection) were then performed by the company-owned waste disposal undertaking whose staff members belong to the closer circle of the 'ECO-team'.

Small steps were initially taken, such as buying ecological detergents in waste-saving refill packaging, and using pump action sprays instead of aerosols. The first step to waste minimisation was ecological purchasing, which did not prove to be more expensive as savings were made in the long run.

A highly motivating factor is the strong involvement of the media, with airport employees being publicly rewarded for their environmental merits. Vienna Airport has its own internal newspaper, which was used extensively to provide information and further encourage the staff. Environmental barometers were placed around the airport which either laughed or cried.

A reward system was introduced to further motivate staff within the airport. An 'ECO-Oscar' was awarded to the department with the most environmentally friendly practices. This has become an annual environmental award and is a very popular trophy, proved by the results of an anonymous opinion poll (with 1000 analysed questionnaires): 96% of all employees were enthusiastic about their company's environmental activities.

The recycling programme has been extremely successful as illustrated in Figs 2, 3, and 4.

This, and the waste minimisation programme has led to a marked decrease of residual waste from commercial activities within the airport, as shown in Fig. 5.

The outline of the project is illustrated in Box 1.

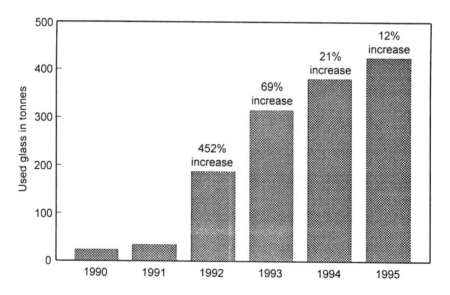

Fig. 2. Used glass collected for recycling

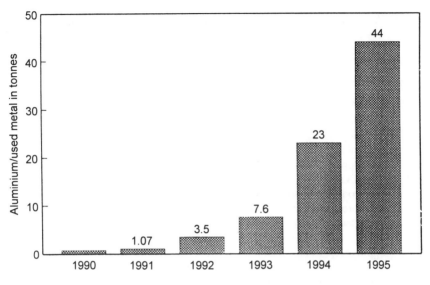

Fig. 3. Aluminium and other metals collected for recycling

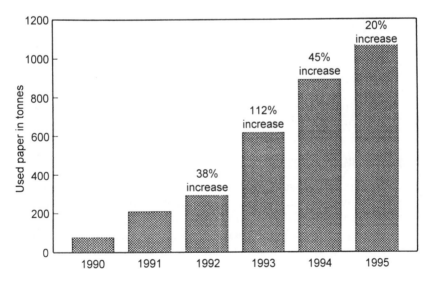

Fig. 4. Waste paper collected for recycling

Box 1: The Vienna International Airport Environmental Protection Model

THE VIE ENVIRONMENTAL PROTECTION MODEL

Principles:

1. We take environmental protection into account in laying down our company objectives.
2. In all decisions taken at VIE, we take into consideration the corresponding ecological aspects.
3. We support environmental measures that promise good results for the overall economy.
   In introducing the measures necessary to achieve this goal, we bear in mind profitability and competitiveness.
4. We answer the call to protect our environment hand-in-hand with the public sector, industry, science and political groups, but above all with our employees, customers (airlines, forwarding agencies, various tenants, etc.), government authorities and institutions (Federal Civil Aviation Authority), as well as our neighbours.
5. Our training courses are programmed taking into account current environmental protection topics.
6. We declare ourselves in favour of progressive legislation.
7. Through comprehensive information we orient our employees, flight passengers, visitors, customers, neighbours, and the general public.
   We realise that due to the high goals we have set, conflicting objectives will repeatedly arise in individual cases. In each case an appropriate solution is to be found.

Some of the specific environmentally friendly measures introduced at the airport are outlined below.

Water pollution

For the last decade, all sewage at the airport has been purified in a fully biological filter plant and reaches the River Danube clean. Constant testing of the water quality in the on-site laboratory guarantees above average purity levels, which was confirmed in a recent environmental study conducted by the Technical University of Vienna.

Energy

Energy saving measures are another focal point of Flughafen Wien AG's environmental programme. Despite the constantly growing passenger volume, effective economising measures implemented in all

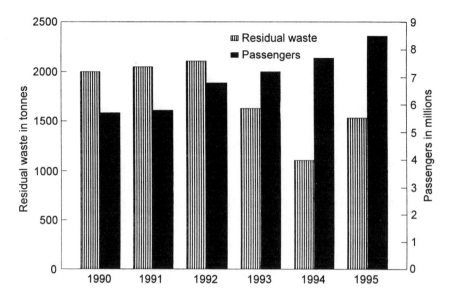

Fig. 5. Residual commercial waste in tonnes

areas of the company have made it possible to save enough energy to supply 500 single-family homes for an entire year.

This was made possible not only by efficient heat recovery systems in several buildings, the use of heat pumps, optimising the piping and thermal insulation, the installation of high-speed gates and energy-saving improvements in all technical facilities, but also by the central process control technology, which monitors all of the power supply and disposal installations. Environmentally-sound construction consultation before each building goes up is the first step toward saving resources at VIE, and all preconditions for the later use of solar energy are taken into consideration at the planning phase. Ongoing tests with electricity-powered vehicles and bio-diesel will determine how soon it will be possible to use optional and environmentally-correct ground vehicles.

The airport technicians are thinking and planning even further ahead: they are designing a pilot project in wind power as a future possibility for saving resources.

Long distance heating

Aware of the energy shortage and that Austria belongs to those countries which consume the lion's share of the world's energy resources,

Vienna International Airport takes advantage of all possible measures to promote economical energy consumption. VIE's heating needs are supplied by a long distance heating transmission line from OMV, a major Austrian Energy company, by utilising existing waste heat. One positive side effect is that there is no direct pollution burden from a heating plant on airport property, which is also reflected in the surprisingly good air quality readings from around the airport.

Fuel pipeline

The pipeline from OMV to the airport also contributes substantially to reducing pollutants: aside from the safety aspect, the pipeline saves our environment some 25,000 trips by fuel tankers annually.

Underfloor tankage system

The underfloor mains feeding turbine fuel to the individual aircraft docking positions is designed exclusively to enhance security and protect the environment. The ATS 270 mn facility does not have any economic advantages.

Winter service

Weather observation and a computer supported weather warning system provide a particularly decisive environmental service in winter. Prompt response to unfavourable ground conditions can reduce the use of ground de-icing agents to a minimum. Generally speaking, VIE applies the environmentally-correct motto: clearing the snow with machines is preferable to using chemicals whenever possible.

Landscaping

Five hundred hectares, approximately half of the surface area of Vienna International Airport, is grassland, which requires landscaping measures. The goal of all greening activities around the airport is to create and maintain natural conditions. In 1992, as part of a major experiment, Flughafen Wien AG tested the impact of magnesium-based organic fertilisation. The results were quite successful.

As the most important element for the green hue of leaves, magnesium not only keeps the airport green far into the autumn, it also protects the soils and groundwater. Magnesium in granular form has several advantages: it can be stored longer; completely prevents penetration into the groundwater, the plant takes only what it needs; and the magnesium makes other soil nutrients available. The magnesium granules that keep the trial areas a deep shade of green until late autumn rate better than comparable synthetic products not only ecologically, but economically as well.

WASTE SEPARATION ON BOARD

While the programme was being introduced in the airport, a one-year pilot project with Austrian Airlines was being run. This proved that waste separation and minimisation is also feasible on board the planes. This was initially implemented without management involvement. Pilots and flight attendants were keen to be involved with the project and began to separate the waste. They replaced plastic bags with linen ones; butter rosettes were introduced in place of plastic tubs; and oil and yoghurt were purchased in larger containers.

Initially this stage of the project was not so successful, as the cost of sorting the waste was higher than incineration. However the majority of the employees were committed to the programme, and the catering management were persuaded to participate.

The next step was to inform all the airlines of forthcoming waste management laws. An incentive system was introduced under which dumping of unsorted waste became very expensive, with a significant surcharge according to the seating capacity. The message that the increasing costs of waste dumps were not simply to be distributed evenly to all parties (through raising costs) but rather to be charged in the form of the surcharge, to those who are not willing to make a contribution, was well received by the airlines (with no higher costs for the environmental activists leading to decreasing costs for the airport). This principle is of course only possible given a truly efficient inspection system.

Since May 1993, every aircraft landing at Vienna Airport has been subject to supervision by authorised environmental 'superintendents'. They check whether waste separation in the sky was actually carried out, inform the crew on board the aircraft about the comprehensive Austrian environmental initiative, and motivate them to join in personally. A 'green airline' award has been introduced to encourage airlines to join the project.

These methods have proved successful: the residual waste quantities had dropped from 15.2 kg per incoming flight in 1990 to 6.6 kg (less than 50%) in 1995 (Fig. 6).

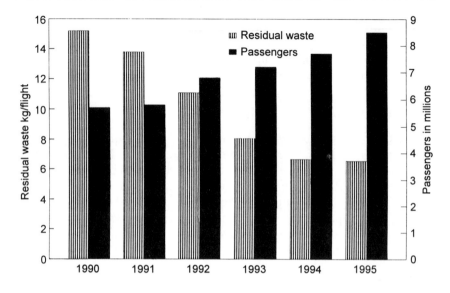

Fig. 6. Residual waste from airlines

PUBLIC PARTICIPATION

The project's last phase, which consisted of motivating passengers, guests, and partner organisations (such as customs, police, fire brigade etc.) to participate in this ECO-program, naturally presented the greatest challenge.

Eye-catching 'islands' were designed, using different colours to represent different materials, to bring the 'waste' problem closer to the 8.5 million passengers (1995 figure) and motivate them to separate their waste. This was not overly successful, as many people are still uninformed about waste management, but providing the information can only lead to an improvement.

Partner organisations were caught up in the enthusiasm of the project and sought ways to improve their own practices. For example, the fire brigade had previously used a granular treatment to absorb hydrocarbon fuel spills, which could be toxic, carcinogenic and corrosive, and when caught by the wind, could prove difficult to collect. A more environmentally friendly method was introduced using a biological clean-up agent, which disintegrates the spill into microscopic droplets, which are neutralised and encapsulated, and can then be naturally biodegraded, with no risk of evaporation into the atmosphere.

CONCLUSIONS

The Vienna International Airport's success story began with providing information, motivation and tailor-made solutions for waste disposal, which triggered a marked increase in the amount of recycling material collected in the air and on the ground.

The message from this project is that environmentally correct behaviour cannot be ordered. By providing the necessary information and by effective communication of this message, by using public and internal relations, the campaign was successful by using human resources rather than by using expensive technology, and in the long run, is cost effective. Each link in the chain must be fully aware of their own importance for only then can the success of the project be assured. This requires comprehensive information, motivation, and inter-disciplinary, intercompany communication, which in this case functions beautifully and with a singular lack of bureaucracy.

Environmental impact assessment: the Norwegian experience

ALICE GAUSTAD, Head of Environmental Affairs Section, Airports and Air, Navigation Services Division, Civil Aviation Administration, PO Box 8124 Dep, Wergenlandsvelen 1, Oslo 0032-1, Norway

The airport serving the Oslo region is approaching its capacity, due to both physical constraints and to environmental impact. Various proposals for a new airport have been made in the past 30 years. A master plan and an environmental impact assessment for a proposed new airport to be located at Gardermoen, north of Oslo, were undertaken in 1991. National Policy Guidelines for planning of the airport, the access system and for regional development, were set out by the Government. The environmental impact assessment was based on these guidelines. Both the content and the extent of the assessment were settled in a program before the studies started. The impact of noise, airport operations and the access system on air and water quality, water balance, natural resources, sites of cultural heritage and socio-economic aspects were evaluated. After a new public inquiry, the Storting (Parliament) made the final decision to build the airport.

The experience gained during the environmental impact assessment process provided a solid basis for a final decision, and a good starting point for an Environmental Implementation Program, to ensure that environmental objectives are implemented both in the design, construction and operation of the new airport. Construction started in 1993 and the new Oslo airport at Gardermoen is due to open in October 1998.

OSLO AIRPORT AT FORNEBU

The main airport currently serving the Oslo region is Fornebu airport, situated 6 km south-west of Oslo city centre. The airport itself is in beautiful surroundings situated by the Oslo Fjord, close to upmarket residential areas. The airport was officially opened as a combined airport for air and seaplanes in 1939. Until then seaplanes formed the major part of air traffic in Norway. At the time, the location seemed ideal, not too far from the city centre, and with few residential areas nearby.

The traffic load at Fornebu has since increased at an amazing rate, in line with many other major airports. In 1960, the annual number of passengers through Fornebu was 600,000. In 1980, this had risen to nearly 4 million passengers. The annual traffic load so far in the nineties is close to 9 million passengers. Fig. 1 illustrates the increase in traffic at Fornebu.

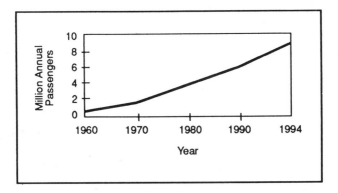

Fig. 1. Traffic increase at Oslo airport at Fornebu

The airport is rapidly approaching its maximum capacity regarding both the number of aircraft operations and its environmental impact.

At peak hours, the number of aircraft movements has reached the maximum capacity, and slots have been introduced. The airport is physically very compact and the potential for better utilisation of the existing facilities is small. Because of the lack of available land, expansion of the runway and taxiway system is not possible. The airport at Fornebu cannot provide the necessary capacity to meet the long term demand for air traffic in the Oslo region.

The other major constraint on Fornebu today is the densely populated surroundings and hence the high number of people that are heavily affected by aircraft noise. Even parts of Oslo city itself are enclosed within the noise contours. In 1990 almost 90,000 inhabitants lived within the noise contours (EFN>55 dBA, EFN is similar to the American CNEL). The introduction of only Chapter 3 aircraft at the turn of the century will somewhat improve the situation but, due to expected traffic growth, the number of people affected by aircraft noise will remain high.

THE HISTORY OF THE NEW OSLO AIRPORT

During the mid-sixties, the Parliament realised that Fornebu could not remain the only civil airport in the Oslo region, and began making plans for the new Oslo airport.

A Governmental committee was set up in 1968 to investigate the need for, and possible location of, a new Oslo airport. This committee concluded that Fornebu should continue as the major Oslo airport, but that excess traffic (mainly charter) had to be transferred to the existing military airport at Gardermoen, approximately 45 km north of Oslo. When these two airports reached their capacity, a new airport would be built at Hobøl, approximately 45 km south-east of Oslo. To secure the Hobøl site the Parliament put an embargo on any other development on the land.

In 1981 new forecasts predicted only 7 million annual passengers at the turn of the century. The embargo was lifted from the land at Hobøl. Air traffic would continue to be divided between Fornebu and Gardermoen.

Traffic increased more rapidly than expected, and in 1984 new forecasts indicated 7 million annual passengers before 1990. There were three choices available: to expand both Fornebu and Gardermoen for continued split traffic, to expand Gardermoen only, or to establish a brand new airport on virgin land. The following sites were evaluated for the latter: Kroer and Hobøl south-east of Oslo, and Hurum southwest of Oslo.

The Parliament decided on the latter site in 1988. A new airport was to be located at Hurum, approximately 45 km south-west of Oslo, and the master planning started that autumn. However, the decision was very controversial. It was based on the following views:

☐ this site would lead to shorter flight distances for the majority of the traffic (approximately 60% of the flights are south – southwestbound, 20% northbound, and 20% east or westbound), thereby reducing the total emissions and flight time

☐ the airport would be closer to the market, which is generally located along the Oslo Fjord

☐ the military activities at Gardermoen would suffer if a civil airport was located at the same site.

Those in favour of Gardermoen emphasised that this airport could be cheaper, as much of the land had already been developed as an airport.

An airport at Gardermoen would promote business and industry in this region, which had been rather slow compared to the region along the Oslo Fjord.

In 1989 the result from continuous weather monitoring showed that crosswinds, combined with poor visibility, made the location at Hurum unsuitable for a new airport. The Parliament then (in 1990) shelved the Hurum proposal and decided that the future Oslo airport would be located at Gardermoen.

The master plan for the airport and access system, based on the preliminary Hurum master plan, was completed in December 1991. After an intense debate outside and inside Parliament the proposal for the construction of the new Oslo Airport at Gardermoen was finally accepted on 8 October 1992.

More detailed planning started immediately, and planning permission was granted in June 1993. Permission for land acquisition was granted in September 1993. Construction works began in August 1993, and the airport is due to open in October 1998. A summary of the history of the plans for the new Oslo airport is shown in Table 1.

MASTER PLAN FOR A NEW AIRPORT AT GARDERMOEN

Organisation (1990–1992)

The Department of Transport and communication had overall responsibility for the planning of the airport and access system. A

Table 1. Summary of the planning history of the new Oslo Airport

Planning the New Oslo Airport	
☐ 1968	Investigation began
☐ 1971	Split traffic between Fornebu and Gardermoen
	Plans for Hobøl to be new Oslo airport
☐ 1984	New forecasts – slow traffic growth
	Plans for new airport shelved
☐ 1984–1988	Considerable increase in air traffic
	New investigations
☐ 1988	New airport at Hurum
☐ 1989–1990	Weather report – Hurum shelved
☐ 1990	Start of master plan for Gardermoen
☐ 1992	Decision to construct new Oslo airport at Gardermoen

Gardermoen-project team was assigned at the Department of Transport and Communication.

The planning itself was carried out by the appropriate public bodies: the Norwegian Civil Aviation Administration were responsible for the actual airport planning, while the Public Roads Administration and Norwegian State Railways were responsible for the access system to the airport.

The Norwegian Department of Defence assessed the consequences for the Military of the use of Gardermoen as a new Oslo airport, taking into account alternative locations for the activities presently based in the immediate vicinity of Gardermoen. The Department of Environment and the Department of Agriculture have co-operated in the assessment of the major regional and social impacts of the Gardermoen. This organisation is illustrated in Fig. 2.

The master planning of the airport was carried out by a separate project group which reported directly to the Technical Director at the Civil Aviation Administration, and which consisted of specialists within the Civil Aviation Administration and consultants who were com- missioned for the task.

The objectives of the master planning that started in 1990 were:

☐   to provide the necessary background information for the impact assessment of locating a new Oslo airport at Gardermoen

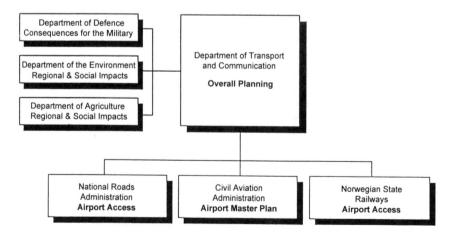

Fig. 2. Organisation during planning (1990–1992)

☐ to plan an airport of high international standard that is efficient, and can meet the demand for air transport in the Oslo region well into the next century.

The main elements in the task are described in Fig. 3.

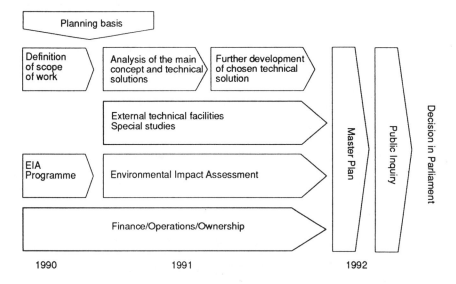

Fig. 3. Main elements of the Civil Aviation Administration's Master Plan (1990–1992)

National policy guidelines

Building a new major airport is of course of general interest to the public. Moving the aviation activity from Fornebu to Gardermoen will entail major changes for the people living in the region of both airports, and we have been through a long planning process with considerable political, professional and public discussions. This has naturally sparked the interest of the media, which may have made the discussion even more colourful.

In recognition of this huge interest, great emphasis has been put on the planning process and achieving the right balance. According to the Norwegian Planning and Construction Act, the Government can set National Policy Guidelines for major developments. Since the establishment of a major airport is of great national and regional significance, the Government found it necessary to state National Policy Guidelines for the planning of the airport, the access system and for regional development. These 'National Policy Guidelines for regional planning and environmental protection for the planning of the new airport at Gardermoen' were instated by Royal Decree on 18 January 1991.

The guidelines have been put forward to secure an open democratic planning process, and to ensure that environmental objectives are included in the planning.

Examples of guidelines and such environmental objectives were:

- ☐ 50% of all passengers are to travel by public transport to and from the airport
- ☐ no major residential development is allowed within the aircraft noise contours
- ☐ directions of regional growth
- ☐ commercial development and settlement patterns.

Guidelines were also set for the utilisation of the Fornebu airport site after the closure of the airport.

Environmental impact assessment

To comply with the Act of Planning and Construction, a comprehensive and detailed Environmental Impact Assessment had to be undertaken before any decision could be made to proceed with the construction stages. The purpose of this Environmental Impact Assessment was to clarify the positive and negative impacts on the environment, natural resources and society. This is to ensure that the impacts are known when a decision is made. Further, it is to ensure that the impacts are taken into account during further planning and in the course of future construction and operation of the airport.

Programme

A programme with the contents and extent of the impact study for the airport was presented by the Civil Aviation Administration. Similar programmes were developed by the Norwegian State Railways and Public Roads Administration for the access system. These were then compiled into one programme for the Environmental Impact Assessment of the entire airport and access system. This programme was sent out to the public and all interested bodies for comments. These comments were then evaluated, and the contents and extent of the Environmental Impact Assessment programme were finally settled by the Department of Transport and Department of Environment in the summer of 1991. Once the Environmental Impact Assessment programme was approved, no further modifications or additions could be made.

The programme included assessments of the airport and access system's impact on:

- ☑ noise
- ☐ air quality
- ☐ natural resources, e.g. water, landscape, minerals, flora, fauna etc
- ☑ cultural heritage sites 文化遺産.
- ☑ socio-economic aspects.

All these issues were to be analysed for two airport alternatives:

- ☐ a new runway to the east of the existing runway
- ☐ a new runway to the west of the existing runway.

The main steps in these studies have been:

(a) investigation and survey of the existing conditions
(b) impact assessment for the two airport alternatives
(c) evaluation of possible measures to avoid/reduce impact
(d) proposal for further studies and monitoring programmes.

Note: this was an impact assessment and no decision as to the appropriate preventative measures was taken at this stage.

The Civil Aviation Administration was responsible for studies on the environmental impact of the airport itself, while the Public Roads Administration and Norwegian State Railways assessed the impact of the access system. Private consulting companies and national institutions were contracted to undertake specialised studies under the programme to cover certain aspects as requested by the responsible bodies. The Department of Environment and the Department of Agriculture have also co-operated in the assessment of the major regional and social impacts of locating an airport at Gardermoen.

Case study example

Cultural heritage was one of the subjects studied in the investigation, and is given as an example of the impact assessment process.

Cultural heritage is divided into Pre- and Post-Reformation (before and after the year 1536). All sites from Pre-Reformation are automatically protected by law.

A. Investigation   The region contains large sites that have great historical value as they are unique records of the earliest history of the area. Especially west of the airport, there are many burial mounds which are traces of an early agricultural settlement. Parts of the area were found to be of special value, as some sites date back to the Stone Age.

The most important Post-Reformation sites were found to be interesting agricultural settlements, particularly at Moreppen. The eastern runway alternative would coincide with the Gardermoen military base which has already influenced the area a great deal since the beginning of this century.

B. Impact   The western runway alternative was found to have a much greater impact on the sites of cultural heritage than the eastern runway alternative. It was considered to be more important to preserve the sites west of the existing runway.

C. Preventative measures   The western runway alternative represents such extensive interference that there did not seem to be any possibility of any modifying or compensating action. For sites close to the airport, area protection and screening would have to be provided, for example through landscaping. It is important that the most valuable sites are preserved in situ. (If they were to be excavated by the appropriate archaeological techniques it would require approximately 350 man-years.)

Measures to be taken for the Post-Reformation sites would be to document/register the valuable buildings, or preferably preservation.

D. Further studies   The western alternative requires extensive archaeological research in order to save some of the historic source material in the ground. The eastern alternative involves far less research.

Studies of this kind were carried out both for the impact of the airport itself and the impact of the access system. Some of the studies, for instance assessing the impact of aircraft noise, were carried out by the Civil Aviation Administration alone, whilst others such as the above study into the impact on cultural heritage, were carried out in collaboration with the National State Railway and Public Roads Administration. The Civil Aviation Administration extracted the relevant information for the actual airport and immediate surroundings.

We also developed what we called 'Theme Maps' for all the subjects in the Environmental Impact Assessment. These were found to be a very useful tool to identify, and easily illustrate, potential conflicts.

As already mentioned, the Civil Aviation Administration did not necessarily have extensive expertise in all these subjects. The additional expertise was therefore brought in from various research institutes and private consulting companies.

POLITICAL DECISION IN PARLIAMENT 1992

Reports on the Environmental Impact Assessment were prepared in late autumn 1991.

When completed, the Environmental Impact Assessment formed a part of the Master Plan for the airport and access system. These were included in a three month public inquiry in December 1991. A considerable number of comments were received from the general public, local communities and from other departments. However, relatively few criticisms were received regarding the depth or scale of the studies. All comments were evaluated. The Department of Transport and Department of Environment concluded that the Environmental Impact Assessment had been carried out according to the programme that was settled in 1991. As a part of this approval, certain conditions and requirements for more detailed planning were set out by the Departments.

Table 2 illustrates the duration of the various stages of planning. Note the large amount of time that was spent on the process itself (much of it to ensure public participation), and the relatively short time spent on the actual studies.

Table 2. Duration of the various planning stages

New Oslo Airport Environmental impact assessment	
1990	Planning started
1991	National Policy Guidelines EIA Programme to the Departments Public inquiry EIA program settled EIA actual studies
1992	Public inquiry EIA approved Parliament decides upon new airport
1993	Oslo Hovedflyplass Ltd est.

The extensive co-operation with local municipalities, counties and various national institutions, has been of great importance. Throughout the investigations and especially the public inquiries, there was considerable feed-back about what is of importance to the public, local, and central authorities concerning the new Oslo airport at Gardermoen. This information was useful when proceeding with planning. The planning process has been very open, in order, hopefully, to avoid potential conflicts at later, more critical stages of planning. This was valuable as it provided a solid platform on which to base a political decision and further planning. The decision to actually build the airport was finally made by Parliament on 8 October 1992.

The following resolution was passed by the Parliament on 8 October 1992:

1. Gardermoen shall be developed as a main airport for the Oslo region, with two parallel runways. The existing runways shall remain and the second runway shall be located to the east of the existing one.
2. A new rail link (the Gardermoen railway) shall be built along the Oslo Central station – Lillestrøm – Gardermoen-Eidsvoll route. The Oslo – Gardermoen section shall be completed in time for the opening of the new airport. The construction of the line between Gardermoen and Eidsvoll shall be in accordance with the Parliamentary Bill.
3. The road E6 (Tangerud – Hvam), RV 174 (Jessheim – Nordmokorset), RV 120 (Erpestad – Gardermoen) road developments and the E6 Arteid bridge shall be completed in time for the opening of the airport.
4. The development of the new airport shall be organised as a public limited company, wholly owned by the state via the Civil Aviation Administration. The state shall transfer its share of the value of the existing airport at Fornebu to this subsidiary (Oslo Hovedflyplass AS).
5. The Gardermoen railway development shall be organised as a public limited company, wholly owned by the State via Norwegian State Railways.
6. Søndenfjeldske Dragoon Regiment (armoured unit), the Special Parachute and Patrol Forces Training Academy and Trandum technical workshops shall be relocated to Rødsmoen in the Åmot municipality.
7. Relocation of the air base, Ecco division, 335 Squadron's B-wing, the Royal Army Service Corps, the military service

refresher course centre and South Gardermoen military camp shall be carried out in accordance with the Parliamentary Bill.

8. The Civil Aviation Administration shall pay NOK 1000 million (M) towards the estimated cost of NOK 2000 M (1992 rates) for the proposed relocation described in paragraphs 6 and 7 above. The remaining NOK 1000 m shall be covered by an increase in the Defence budget. The amount shall be transferred as relocation is carried out.

9. Revenues from the sale of land and buildings belonging to the Norwegian Armed Forces shall go entirely to the Armed Forces.

10. The new airport for the Oslo region shall be an outstanding example of Norwegian architecture. Consequently, an integrated plan for artistic design shall be developed in connection with the architectural work.

PLANNING AFTER THE POLITICAL DECISION (1992–1995)

Organisation

The Civil Aviation Administration established the subsidiary named Oslo Hovedflyplass Ltd in November 1992. This subsidiary was given the responsibility to undertake the detailed planning/design, construction and operation of the new Oslo Airport at Gardermoen in accordance with Parliament's decision.

The Oslo Hovedflyplass Ltd's most immediate tasks were to:

☐ obtain the necessary planning permissions
☐ establish an environmental plan
☐ start the detailed design and construction.

The Civil Aviation Administration's responsibility was to:

☐ ensure that Oslo Hovedflyplass Ltd carried out the above tasks according to Parliament's decision
☐ carry out the necessary land acquisition
☐ establish a new traffic regulation system and airspace organisation for East Norway
☐ preliminary land-use planning for Fornebu.

Revised national policy guidelines and planning permissions

The Department of Environment set out revised National Policy Guidelines for the continuation of planning, construction and operation of the new airport and access system. These were very much in accordance with those set out in 1991, but were now more detailed.

More detailed guidelines were also set out for utilisation of the Fornebu airport site after the closure of the airport.

The use of the land at Fornebu shall be for:

☐ residential use
☐ recreation along the shore
☐ industry, where suitable buildings for this purpose are already located.

Planning permission was granted in June 1993. This was a land-use plan that set the framework for the utilisation of the area. It also gave formal approval for starting the process of land acquisition and for the start of construction. The permission granted covers an area of 13.000 daa, enclosing approximately 270 private properties (area: 3.700 daa). Of these, 230 were within the new airport boundary, and had therefore to be vacated. The State, through the Department of Defence, owned 75% of the area.

The development of Gardermoen as a new Oslo airport is the largest onshore development project in Norwegian history. The ambition is to build 'Europe's most modern airport', at a total cost of NOK 20 billion (1992 rate, including the railway-link) within the specified time and budget, and according to specified environmental requirements. The objective is that the new Oslo airport at Gardermoen, when completed in autumn 1998, will be a supreme exhibition of Norwegian architecture designed to impose minimum impact on the environment.

To achieve this objective, we have to comply with some of the strictest and most ambitious environmental requirements ever imposed on a construction project in Norway.

Environment monitoring programme

At this point it was important to keep in mind and use the lessons learned throughout the Environmental Impact Assessment. While applying for the necessary planning permissions, the Department of Environment required an Environmental Monitoring Programme. This was to ensure that the knowledge gained about the environmental impact will be implemented throughout the planning, detailed design, construction and operation of the new Oslo Airport.

This programme was produced by the Oslo Hovedflyplass Ltd and published in October 1993. It covers all the environmental themes covered in the Environmental Impact Assessment, including land acquisition.

The Environmental Monitoring Programme has been divided into the following sections:

1. environmental objectives
2. implementation of the environmental objectives in detailed design, construction and operation
3. special projects
4. environmental and social surveys, and monitoring.

The Environmental Monitoring Programme was made available to relevant bodies for comments, and was thereafter formally approved by the Civil Aviation Administration, in co-operation with the Department of Transport and Communication and Department of Environment.

Environmental monitoring is not a static programme. It is a process that takes place throughout the entire project. It concentrates on establishing relevant environmental objectives/principles within the project, which also applies to external consultants/contractors. In addition, the programme describes how these environmental objectives are to be implemented in the contracts. Thus, the program will ensure that environmental implementation is a process where measures/mitigation are established underway.

Environmental objectives

The Civil Aviation Administration and Oslo Hovedflyplass AS have based all future work with the new airport on the following environmental policy:

☐   the new airport shall be adapted to the environment, and will be developed and operated in a way that minimises the negative environmental impact
☐   environmental considerations shall be integrated in the planning, construction and operation of the new airport on a par with functional, technical and financial considerations.

Primary objectives have been established for all environmental elements covered in the Environmental Impact Assessment, plus the land acquisition. More detailed conditions (requirements for solutions) have also been established. These are to cover all the requirements laid down by legislation, regulations and conditions given in the approval of the Environmental Impact Assessment.

Below is an example of how the objectives for cultural heritage are stated. This example is chosen to follow on from the previous example given for the Environmental Impact Assessment.

An example of environmental objectives The airport will have an impact on an area with important archaeological remains ranging from the early Stone Age to more recent times. Pre-Reformation (before the year 1536) cultural heritage sites are automatically protected according to the Cultural Heritage Act. Archaeological findings include tools, burial mounds and evidence of grazing/early farming. A burial mound has been registered at Vigsteinsmoen, which borders the airport. The area may also contain remains from the Iron Age. In addition, evidence of the use of coal, dating from 100 BC to 1200 AD, has been registered in the eastern part of the new airport, and to a greater extent outside the new airport boundary. From more recent times, the military installations at Gardermoen are regarded as cultural monuments.

The cultural monuments in this area are primarily of regional importance. The Gardermoen military camp and civilian buildings around the parade ground have to be removed.

Primary environmental objective for Cultural Heritage The historical development of the area, from the first settlements to present day, shall be documented, and particularly important cultural monuments and historical structures shall be protected. Construction works shall be carried out in such a way as to present minimum risk to cultural heritage.

Principles – mitigation measures A separate plan shall be established for the administration of cultural monuments in the airport area. The plan will ensure the protection of a number of cultural monuments and structures as evidence of the historical development of the region. Buildings that are of particular importance, and which do not fit in with the airport development, shall be moved to another site with the possibility of being used as museum. The plan will also determine how historical remains in the airport area are to be recorded and administered in order to supplement and explain the physical cultural monuments that will remain. The historically important Vigsteinsmoen, west of the western runway, has been set apart as a special area of protection in the Planning Permission.

Implementation of the environmental objectives

The objectives and conditions described above are to be implemented in all the activities that have an environmental impact. Hence it forms the basis for pre-qualification of consultants/contractors, tendering and contracts between the developer and suppliers, and it is linked to individual activities. It also covers the procedures for subsequent controls and audits.

Of course, the airport development comprises many elements. Each of these elements will affect the various environmental aspects to varying degrees. Therefore, establishing the criteria will also vary. To ensure that future work focuses on the most important environmental aspects, the impacts of various activities have been prioritised.

The matrix below (Fig. 4) illustrates the correlation between the various activities that will take place and their impact on the environmental aspects. The darkness of the cells indicates the level of importance, hence black means 'grave impact' and white 'no significant impact'.

The environmental objectives are to be considered as permanent, while the environmental principles and procedures for monitoring and control will be evaluated regularly, with a view to making any necessary adjustments.

After the environmental objectives/principles have been integrated into the contracts, the contractor must follow up with documentation and control of technical solutions and design. Procedures shall be established, and these shall be a part of the contract and linked to fixed milestones. They will eventually be integrated into the quality assurance system. The procedures comprise requirements for control by the contractor, documentation of results and reporting to the developer. They also deal with procedures for managing non-conformities between requirements and results, which may lead to reappraisal of the requirements, and with communication procedures and emergency measures in the case of accidents during construction.

Special environmental projects

For some of the subjects, it has been necessary to undertake further studies in order to establish conditions for the desirable level and possible mitigation measures. Examples include a comprehensive plan to avoid pollution of the groundwater reservoir and the planning of the

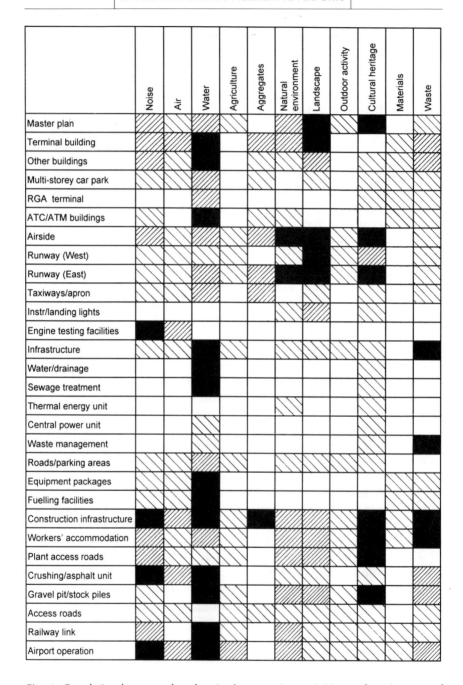

Fig. 4. Correlation between the planning/construction activities and environmental impact

new flight paths to and from the airport. The latter example can be used to demonstrate how this work with the flight paths is described in the Environmental Monitoring Programme.

Background and objectives Procedures for landing and take-off at the new airport at Gardermoen will be established by the Civil Aviation Administration as the aviation authority. The procedures are stated in regulations according to the Act of Aviation.

On approval of the Environmental Impact Assessment, The Department of Transport and Communication stated the following conditions while approving the Environmental Impact Assessment:

> The establishment of inbound and outbound flight paths must be evaluated in the light of capacity and safety aspects as well as environmental aspects. The final pattern of the flight paths shall be decided at a later stage. This will be based on airspace simulation tests. These simulations will be based on, amongst other things, a prioritising of areas that are defined as noise-sensitive areas. The Civil Aviation Administration, in consultation with local, and other authorities will initiate a programme to identify these noise-sensitive areas.

Schedule The process of identifying these priority areas was initiated in 1993. In 1994 the Civil Aviation Administration and Oslo Hovedflyplass Ltd will develop the air traffic regulation system. The air traffic regulation system will be simulated in the winter of 1994/95. The plan will be compiled in 1995. Proposals for regulation will be sent for a public enquiry, in accordance with the Public Administration Act, before the new airport is commissioned.

Co-operation Oslo Hovedflyplass Ltd has entered into co-operation with the four municipalities nearest the airport. These will suffer future land-use restrictions due to noise. This will be discussed in detail by an aircraft noise committee for Gardermoen, set up by the Government. The four municipalities, Ullenasaker, Nannestad, Gjerdrum and Eidsvoll, will be represented in this committee.

Oslo Hovedflyplass Ltd has also entered into co-operation with the municipalities of Oslo, Hurdal, Skedsmo, Nittedal, Sørum, Fet, Rælingen and Lørenskog, as well as Akershus County. Although these municipalities are situated outside the noise zones, they will nevertheless be affected by increased air traffic to and from Gardermoen.

Such 'special projects' that run parallel to the general planning and construction were set up for:

☐ Cultural Heritage Plan (as mentioned under Environmental objectives above)

☐ establishing flight paths (as mentioned above)

☐ water resources
  – analysis of run-off water and snow melt from the airport area
  – drainage of waste water from the new airport
  – protection of groundwater and ground filtration characteristics
  – lowering the groundwater table for the construction of the railway culvert and station

☐ evaluation of environmental impact

☐ landscape plan

☐ waste treatment during construction

☐ provision of property/land as compensation for appropriated property

☐ odours and oil deposit

☐ evaluation of the possibilities for screening of aircraft noise.

Environmental monitoring and social surveys

A separate programme for environmental monitoring has been developed. This program will provide information that will help monitor the effect of the airport on noise, air quality, water resources and the natural environment (flora and fauna).

When stating the licensing conditions, the environmental authorities will specify requirements for environmental monitoring of the individual activities for which permission will be granted.

The aim of the monitoring is:

1. to ensure that recommended pollution limits given in relevant legislation, guidelines and licences, or limits proposed in planning (the Environmental Impact Assessment), are not exceeded: any indications that limits are exceeded shall be reported and preventative measures considered

2. to provide information from the surveys as a part of a database that will be used for documentation of possible environmental changes as a result of the development and operation of a new airport at Gardermoen.

In addition, a separate programme is set up to study the social effects on the people living at and near the airport. The latter will be undertaken by the local authorities with Oslo Hovedflyplass Ltd as a participant.

The detailed design and construction that has been carried out so far, has been in accordance with this Environmental Monitoring Programme. Environmental monitoring is a continuous process throughout the project. The Environmental Monitoring Programme may require adjusting along the way, depending on how well the established principles and monitoring procedures work.

THE NEW OSLO AIRPORT AT GARDERMOEN

So far, the detailed design and construction is proceeding as planned. Fig. 5 illustrates the schedule for the main activities.

During the planning process several alternative plans of the layout of the new airport at Gardermoen were assessed, but the final choice is based on the objectives of obtaining optimal operational conditions, maximum flexibility, minimum ground movements and, of course minimal environmental impacts. The development plan is based upon the assumption that the existing airport must be in operation throughout the construction period, without any interference from the construction work.

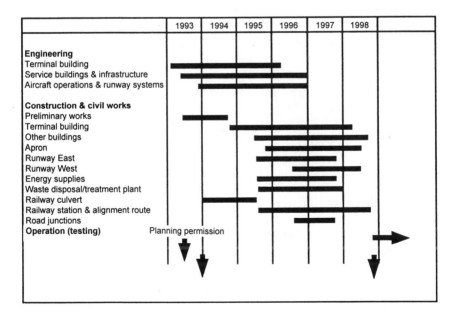

Fig. 5. Schedule for the main activities

The airport is designed to handle 12 million passengers in the year 2000. This can be increased to provide for approximately 17 million passengers annually, before any major expansion is necessary. The runways are spaced at 2200 m to provide independent operations and provide sufficient building area for the various functions and technical structures at the airport. The runways are staggered to achieve the most efficient ground movements.

The terminal building configuration is based upon a central terminal with a concourse and a satellite parallel to the concourse located 340 m from the main building. The satellite is connected to the main building by an underground moving walkway. Land for a second satellite has been provided. Forty gates with loading bridges plus fifteen hard stands will be provided at the time of the opening of the airport.

As mentioned, one of the objectives in the National Planning Guidelines was that 50% of the passengers are to access the airport by public transport. To achieve this the principal means of public transport to and from the airport will be by rail. A railway station is located underneath the terminal, with easy access to the terminal building by elevators and escalators. The rail link Oslo – Gardermoen will have tailor-made carriages, high departure frequency and a travel time of an estimated 19 minutes.

The airport will be financed by income from the sale of Fornebu and equity from the Civil Aviation Administration, and a loan of approximately 8 billion NOK. The economic analysis shows that the project will give a rate of return of 7% and that the project is economically viable.

EXPERIENCE SO FAR

Planning is not only to create a plan. Planning an airport is to take part in and conduct a process that involves a complex network of public and private bodies and interests. This might be obvious to the professional planner but it may not always be so simple for the public and local communities to relate to. The importance of participation of the public can never be emphasised enough ('participation for pacification').

Environment

Other authorities (State Pollution Control, County Governor etc.) have given the Civil Aviation Administration (and Oslo Hovedflyplass Ltd) as the developer, increasingly strict environmental requirements. These may be stricter than what would be expected according to the existing laws and may even establish a precedent for future developments, not only within aviation. The authorities may be using this large project as a tool to establish a precedent for stricter environmental requirements, instead of changing the general law and guidelines that have applied so far. Aircraft noise and water pollution have been the topics most focused on.

The processes involving public enquiries and obtaining the various necessary permission is time consuming and needs a lot of follow-up from the developers. As mentioned, aircraft noise has been particularly focused upon. The municipalities from Oslo (south) to Eidsvoll and Hurdal (north) have participated. This was to prioritise noise sensitive areas over which there should be no direct flights. It has been difficult to achieve this, as there are many conflicting interests from municipality to municipality. An early public enquiry for the proposed flight paths demonstrated that there is interest in the flight paths, even in areas outside the noise contours. There are conflicting interests between residential areas and recreational areas.

Land acquisition

The focus here has been on the law, the process and social relations. The Government has based the land acquisition on the existing laws and practices. Many, especially local politicians, have disputed this. Their view has been that this is such a special case in Norway (never since the Second World War have so many people been forced to move) that the home owners should be granted some special compensation in addition to the compensation they are entitled to according to the law.

The process has also been criticised. This is largely because some people wanted to move and collect the compensation immediately after the Parliament's decision in 1992, whilst others wished to wait as long as possible. This has been complicated to handle, and has also been difficult for those staying, as it has left large parts of the residential area vacant, thereby making the neighbourhood even less attractive. The media has been very interested in this process, and have followed up with stories about unhappy and scared people fighting the authorities while living in a 'ghost village'. The municipality (with financial support from the Civil Aviation Administration) has supported these people with advice and practical help with moving.

The law

Such major developments as the construction of a new airport are to be planned according to the Act of Planning and Construction. This Act acts as an 'umbrella law' and shall also comprise environmental considerations (the requirement of an Environmental Impact Assessment is according to this Act). The flight paths will be stated according to the Act of Aviation.

However, experience shows that quite a few state bodies try to use 'their laws' to control the process. An example is the local health authorities who have tried to use the 'Act of Health in the Municipalities' to make Oslo Hovedflyplass Ltd undertake more extensive studies before a final recommendation for the new flight paths can be proposed. This law has also been used to set out stricter requirements for the Oslo Hovedflyplass Ltd to treat surface water.

It is therefore necessary that the Government sectors involved decide whether or not one Act can overrule another. If this is the case, it is almost impossible for the developers to know in advance which Act to base the planning upon.

The organisation

It has been both useful and necessary to have the special Gardermoen team in the Department of Transport and Communication, which has co-ordinated the work between the various departments.

There has been a need for extensive legal expertise, both for the developers and the authorities, throughout planning and construction.

The establishment of Oslo Hovedflyplass Ltd with its designated tasks (detailed design, construction and later operation of the airport) has so far been considered successful.

Political interest

There has been – and is – a great interest from the politicians in this project. Of particular interest have been:

☐ land acquisition
☐ the decision of which alternatives to choose for the railway line from Oslo to Gardermoen. The debate about compensation to those living near the railway line and the debate about a tunnel in the most densely populated areas have been focused on

☐ the Environmental Monitoring Programme, including aircraft noise and protection of the groundwater reservoir.

Long-term planning

It is important to stand firm on long-term traffic forecasts. Forecasts should not change on the basis of annual traffic fluctuations even if there is a public demand to do that. To comply with such demands will lead to a very expensive stop-and-go planning process. It is very difficult to keep the politicians' and the public's attention and understanding of a long-term need. Therefore, investment in information about why it is necessary to increase the airport capacity in the future, and the consequences if the future traffic demand is not met, is of importance.

Environment management within the Federal Airports Corporation (Australia)

DESIREE LAMMERTS, Chief Environment Coordinator, Federal Airports Corporation, Locked Bay No 28, Botany, NSW 2019, Australia

The Federal Airports Corporation (FAC) Australia, owns and operates twenty-two airports throughout Australia. In 1993 the FAC completed its first comprehensive environmental audit and the Board endorsed a Corporate Environment Management System.

The system, which is based on the British Standard 7750, Environment Management Systems, is currently being implemented at all FAC Airports. It consists of a Corporate Environment Policy, a statement of Corporate Environment Responsibilities, and an Environment Management Manual.

While the Environment Policy and the Corporate Environment Responsibilities are generic, the Environment Management Manual provides a performance specification for the development and implementation of the Environment Management System within each airport. It allows individual airports the freedom to develop an Environment Management System tailored to their needs, while ensuring a consistent environment management standard throughout the Corporation.

This paper deals with the development of the FAC Environment Management System within the FAC, the reasons for its adoption and the benefits it has for management, the airports and the community. The paper also considers the future goals for environment management within the Corporation.

INTRODUCTION

The Federal Airports Corporation (FAC) is a government business enterprise, established by the Federal Airports Corporation Act (1986).

The FAC owns and operates 22 airports geographically distributed throughout Australia (Fig. 1). For convenience these airports are split into one of the following three classifications.

Fig. 1. Location of FAC Airports in Australia

☐ Primary airports (generally the larger airports)
  – Sydney
  – Melbourne
  – Brisbane
  – Perth
  – Adelaide
  – Tasmania (Hobart and Launceston)

☐ Regional airports (generally the medium-size airports)
  – Coolangatta
  – Canberra
  – Darwin
  – Townsville
  – Mount Isa
  – Alice Springs
  – Tennant Creek

☐ General aviation airports (generally the smaller airports)
  – Essendon
  – Bankstown
  – Hoxton Park

- Camden
- Moorabbin
- Parafield
- Archerfield
- Jandakot.

In the 1993/94 financial year the FAC's revenue was in the order of $457 million. In that time its airports handled over 49 million passengers and landed some 26 million tonnes.

Annual revenue at individual airports ranged from $0.8 million to $174.5 million, while total aircraft movements ranged from roughly 9000 to 650,000 and airport staff strength from 2 to 345.

In order to manage this complex network of airports, the FAC has established a management structure as shown in Fig. 2. The Corporate Head Office is located in Sydney, Australia's largest city and home of the airport with the largest revenue (Sydney Airport). There are approximately 80 head office staff.

Based within Head Office is the Chief General Manager (Operations) who is responsible for reporting all airport related matters to the Managing Director.

The General Managers of the Primary Airports report directly to the Chief General Manager (Operations). In the case of the Regional Airports and the General Aviation Airports however, the Airport General Managers, and Airport Managers report respectively to the General Manager (Operations and Regional Airports) and the General Manager (General Aviation), both of whom are based in Head Office. These General Managers in turn report to the Chief General Manager (Operations).

In addition, the Managing Director has a number of other senior managers based in Head Office, to provide advice and support on matters such as finance and planning, legal, commercial, technical and construction, human resources and corporate affairs.

Environmental matters, including the development and implementation of the FAC Environment Policy and Management System, fall within the responsibility of the General Manager (Technical & Construction) and are managed by the Chief Environment Coordinator.

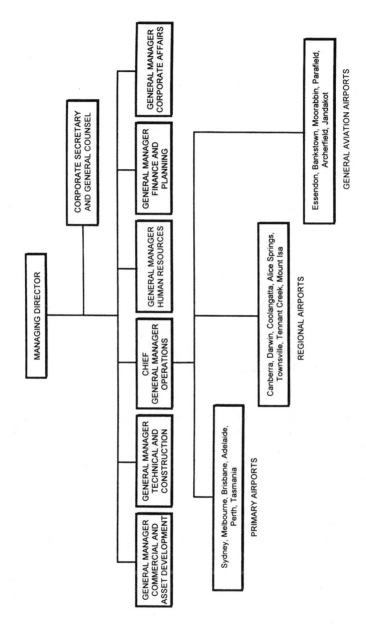

Fig. 2. Federal Airports Corporation Management Structure

ENVIRONMENTAL DUE DILIGENCE

In Australia, the enacting of Federal and State environmental protection legislation, especially in the last decade, has led to legal mechanisms being established to enable directors and company officers to be held personally accountable for failure to meet the requirements of the legislation unless they are able to establish one of the available defences.

The legislation dealing with penalties is as follows:

New South Wales	– Environmental Offences and Penalties Act 1989
Victoria	– Environment Protection Act 1990
South Australia	– Environment Protection Act 1993
Queensland	– Environment Protection Act 1994
Tasmania	– Environment Management and Pollution Control Act 1994

Corporate fines of up to $1,000,000(AUD), and in the case of Queensland up to $1,250,000(AUD), may be imposed and individuals may be subject to penalties of up to $250,000(AUD) and/or imprisonment.

Environmental management therefore requires all those charged with the responsibility to manage businesses to have a clear understanding of, and to carry out, their specific environmental responsibilities.

Most environmental legislation in Australia provides for the defence of Due Diligence, as a means of either preventing conviction or reducing the penalties applied as a result of prosecution for non-compliance with environmental legislation.

The concept of due diligence has been defined to some degree by the following unreported decision of Mr. Justice Hemmings in the *State Pollution Control Commission (SPCC)* v. *Kelly* (Hemmings J, No. 50190, 26 June 1991). In this case, the Director of a waste disposal company was convicted of an offence under the NSW Environmental Offences and Penalties Act 1989, after the company was found guilty of breaching the Act. The Director pleaded not guilty and argued that he had exercised due diligence to prevent the contravention by the company. The Court rejected the plea.

In his decision Justice Hemmings stated,

'Due diligence, of course, depends upon the circumstances of the case, but contemplates a mind concentrated on the likely risks. The requirements are not satisfied by precautions merely as a general matter in the business of the corporation, unless also designed to prevent the contravention'.

In the recent Canadian decision of the *Queen* v. *Bata Industries Ltd.* (unreported, 7 February 1992, Provincial Offences Court of Ontario, Canada) the Court identified that the following questions should be addressed in order to assess the due diligence of directors.

☐ Did the Board of Directors establish a pollution prevention system? [That is], was there supervision or inspection? Did they exhort those they controlled or influenced?
☐ Did each Director ensure that the Corporate Officers have been instructed to set up a system sufficient within the terms and practices of its industry of ensuring compliance of environmental laws, ensure that the officers report back periodically to the Board on the operation of the system, and ensure that the officers are instructed to report any substantial amount of non-compliance to the board in a timely manner?

This case clearly establishes that in order to demonstrate due diligence directors must be able to establish the following:

☐ they instructed relevant officers to establish a Pollution Prevention System
☐ they ensured that the system established was suitable and sufficient to ensure compliance with the environmental laws applying to their industry
☐ they instigated regular checks to ensure the system was operating effectively
☐ they received regular reports on the compliance of the system and ensured that non-compliance was reported in time for remedial action to be undertaken
☐ they remained aware of the environmental standards applying to their industry, and
☐ they personally reacted to rectify any failure within the system.

Within the FAC, a system has been developed to enable the Corporation and its officers to conduct their affairs consistent with the above and thereby to demonstrate due diligence.

The following sections outline this system and its implementation.

DEVELOPMENT OF THE FAC ENVIRONMENTAL MANAGEMENT SYSTEM

Introduction

A key stimulus to the development of the Environmental Management System was as a result of a recognition by the members of the Board of their personal responsibilities and accountabilities under environmental legislation, and the potential negative financial impact on the Corporation of poor environmental performance.

The first major initiative taken by the Corporation was to undertake audits at all 22 airports in order to establish the nature and extent of environmental issues requiring attention. Since this is such a major undertaking, the airports were assisted in this process by the production of 'Guidelines for Conducting a Review of Environmental Management Practices on FAC Airports'.

Staff from all airports attended a two-day training session, run by the FAC Chief Environment Coordinator, to familiarise themselves with the requirements of the audit process. A follow-up meeting halfway through the audit process also was held. These meetings were invaluable in assisting with the resolution of important practical matters, which could have impeded the review process, including:

☐ who should conduct the review (level of expertise required)
☐ the issue of confidentiality
☐ how to deal with tenants
☐ the applicability of particular State and Federal Government legislation
☐ how to handle the information generated by the review
☐ how to report results.

Experience with the review process revealed that a framework for effective management of environmental matters within the FAC was necessary, and the FAC commenced development of this framework. The Environmental Management System which the FAC developed was based on the British Standard 7750 and draft Australian Standards 94376 and 94377 (based on ISO 14000 and ISO 14001).

Extensive consultation was undertaken with the Airport Managers during the development of the FAC Environmental Management System, and this has been vital in encouraging acceptance of the system and to facilitate its implementation at the Airport level.

Dimensions

One of the key aspects of the system is the requirement that the responsibilities be implemented through the current line management responsibility delegations.

A feature of the FAC organisation is the pivotal role of the Airport Manager. Each airport is operated as a largely autonomous business unit, managed by the Airport Manager.

The role of the Managing Director and Corporate Head Office is to establish the policies of the Corporation (with the approval of the FAC Board), to support the Airport Managers in carrying out these policies, and to monitor and control the system so as to ensure it is effective.

In this context, the FAC Environmental Management System may be viewed as having a 'vertical' and a 'horizontal' dimension (Fig. 3).

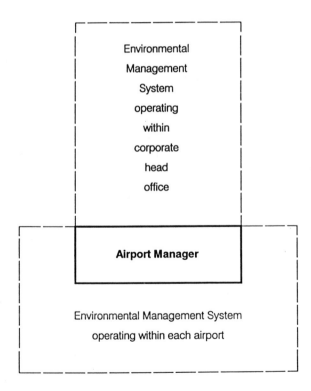

Fig. 3. Relationship between Environmental Management Systems operating within each individual airport and corporate head office

The 'vertical dimension'

The vertical dimension includes the responsibilities and actions to be taken by the Corporate Head Office staff and the FAC Board. It also includes the specific responsibilities of the Airport Managers (Fig. 4).

The components of the vertical dimension are:

- ☐ FAC Environmental Policy
- ☐ policy implementation statements including
  - – allocation of responsibilities, accountabilities and authorities to the Board, Managing Director, Head Office Staff and Airport Managers
  - – amendments to position descriptions of above
  - – amendments to performance appraisal criteria of above
- ☐ Board Environmental Committee
- ☐ Environment Review Group
- ☐ routine reporting requirements
- ☐ incident reporting requirements
- ☐ review requirements.

The Board Environmental Committee is a sub-committee of the Board consisting of the Managing Director and two other Board members and has the following terms of reference:

- ☐ To review and make recommendations to the Board on
  - – the development and revision of FAC Environment Policies
  - – the resources required to implement environment policies
  - – the priorities for capital expenditure required to remedy non-compliance with environmental standards on FAC airports.
- ☐ To ensure that the Corporation acts with due diligence in relation to environmental management by monitoring and reporting to the Board on the performance of the Environment Management System.

The Environment Review Group, on the other hand, consists of those Managers within Head Office with direct responsibility for the management of environmental issues throughout the Corporation. Membership consists of the General Manager – Technical & Construction, Chief General Manager – Operations, General Manager – Finance & Planning and General Counsel.

The function of the Environment Review Group is

- ☐ to review the FAC Environmental Policy at least every two years and to recommend changes to the Managing Director

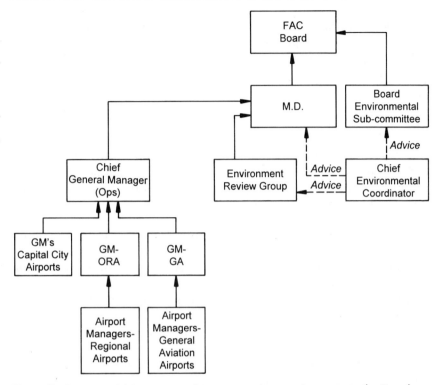

Fig. 4. Environmental Management System reporting requirements to the Board

☐ to monitor trends in environmental legislation, Government policy and initiatives, and standards, and to recommend to the Managing Director changes required to the Airport Environmental Management Systems

☐ to review the monthly environmental performance reports from the airports and to prepare a report to the Managing Director evaluating trends in environmental performance

☐ to review the priorities for action arising from the monthly environmental performance reports, and to recommend to the Managing Director the appropriate use of FAC resources to achieve the aims of the Environment Policy

☐ to conduct a regular management audit of the implementation of individual Airport Environment Management Plans, and to advise the Managing Director of the status of implementation

☐ to review the operation of the FAC Environmental Management System every two years, and to recommend changes to the Managing Director

☐ to investigate and develop proposals for the Managing Director regarding environmental insurance.

In summary, the vertical dimensions of the FAC Environmental Management System provide the framework in which the FAC Board and Corporate Head Office can demonstrate due diligence in carrying out their responsibilities under environmental legislation.

The 'horizontal dimension'

The horizontal dimension of the FAC Environmental Management System is, in effect, a mini-system applicable to each airport. This is essential given the relative autonomy of each of the 22 airports which form the Federal Airports Corporation.

To assist each airport to implement a suitable system, and to encourage uniformity in reporting to Corporate Head Office, a guideline document was developed, referred to as the Airport Environmental Management Manual (AEMM). This document specifies the elements each airport must develop and implement to establish a functioning environment management system, known within the Corporation as the Airport Environment Management Plan (AEMP).

The major elements which the AEMM specifies must be incorporated within each AEMP are outlined below.

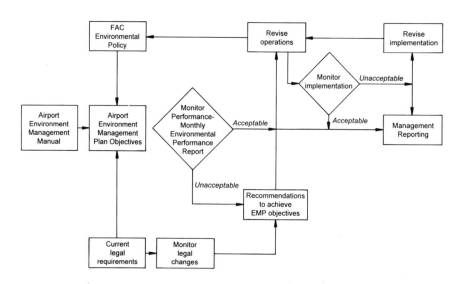

Fig. 5. Operation of FAC Environmental Management Plan within each airport

Management responsibility and organisation

(a) *Responsibility within the FAC.* Responsibilities for implementation of the FAC Environmental Policy are to be clearly defined within each airport. While the primary responsibility for the development and operation of the AEMP rests with the Airport Manager, responsibility to implement and maintain the Plan may be delegated to an Airport Environment Manager. (In practice this role has been delegated to the Technical Services or the Operations Manager at most airports. At some Airports there is a dedicated Environmental Officer to assist the Airport Environment Manager in this task.)

(b) *Responsibilities of tenants.* The environment responsibilities of tenants are the same as those of the FAC. The AEMM recommends that each airport institute an Airport Environment Committee, consisting of FAC officers and tenant representatives, to facilitate cooperation in managing common environmental issues on the airport.

(c) *Responsibilities of contractors.* The AEMM requires the environment responsibilities of contractors to be specified within their contact conditions, and failure to meet these conditions will constitute a breach of contract.

Awareness and training

Successful implementation of any policy requires training, and the AEMM specifies it will take three forms:

- ☐ awareness training, which will also form part of normal induction training
- ☐ training in environmental procedures for managers, technical personnel and relevant employees and
- ☐ motivational training, to encourage all employees to have a regard for environmental matters.

Management reviews

The success of any management system depends on regular management reviews of the effectiveness of the system.

The AEMM requires each airport to establish a schedule of internal audits of all elements of the AEMPs at not greater than twelve-monthly intervals.

Environmental management objectives

While the FAC Environment Policy applies across the Corporation, each airport is encouraged to develop airport specific environment management objectives to suit their unique environments and individual management styles.

Specific management objectives and performance targets are, however, required within the Environment Action Plans for identified environmental effects (see below).

Records

Each airport is required to establish a secure record system including:

- [ ] the AEMM and the AEMP
- [ ] reports of audits and reviews of the AEMP
- [ ] reports of non-conformance and 'corrective action taken'
- [ ] reports of any incidents, complaints and follow-up actions taken
- [ ] monitoring data
- [ ] calibration records
- [ ] records of training and inductions
- [ ] records of permits or licences.

Routine reporting requirements

The AEMM specifies that the Airport Manager is required to provide a monthly written performance report to the Chief General Manager (operations), and a copy to the Chief Environment Coordinator. This management report will detail the status of implementation of the system, and actual performance against targets.

Environmental Accident/Incident Reporting requirements

Prevention of adverse environmental effects is one of the principal aims of the FAC Environmental Policy, and to this end the AEMM requires that all employees and contractors are required and encouraged to report any accidents/incidents or near misses. A standard Accident/Incident Reporting Procedure has been developed, and all employees and contractors are made aware of the appropriate procedures at induction.

It is a requirement that if the accident/incident is likely to have significant adverse environmental impacts, or to lead to the involvement of regulatory authorities and/or negative public reaction, the Airport Manager must report the accident/incident to the Managing Director and the General Manager Technical and Construction.

Consultation with regulatory authorities and the community

Each airport is required by the AEMM to establish formal procedures to ensure that consultation is managed, and does not occur in an uncontrolled and therefore inappropriate manner.

Contractors

Each airport is required by the AEMM to review its standard contracts for the engagement of contractors, to ensure that all contractors undertaking works on the airport are required contractually to comply with the FAC Environment Policy, and to comply with all environmental legislation and standards relevant to the works being undertaken.

Tenants

It is the policy of the FAC to encourage its tenants to adopt environmental policies consistent with the FAC Environmental Policy. The FAC has reviewed its standard tenancy lease documents, and each airport is required to review existing leases with tenants, in order to meet this objective. In addition, by-laws are being developed to require tenants to meet their environmental obligations.

Complaint handling procedure

A standard procedure has been developed for dealing with environmental-related complaints.

Purchase and disposal of major (property) assets

The AEMM requires that the airport obtain a specialist environmental assessment of any proposed (property) asset purchase or transfer, before decisions are made to proceed. The FAC Property Manual has been amended to reflect this requirement.

Purchase and disposal of plant and substances

The AEMM requires the institution of a purchasing policy for plant and substances, to ensure an evaluation of potential adverse environmental effects and to encourage the use of 'environmentally-friendly' alternatives.

Similarly the AEMM requires that the disposal of plant and substances only occurs after an assessment has been made of any potential adverse environmental effects.

Environmental impact of major developments

The AEMM requires that each airport follows the procedures already established in FAC Technical Memorandum 1.4.001 (Environmental Assessment of Proposals), to ensure compliance with the Federal Environment Impact Assessment legislation.

Environmental Action Plans

This is the part of the system which details the specific environmental objectives and targets of the airport for identified environmental effects, and the means by which they are to be achieved.

The plan is to be structured to address each potential environmental effect by way of an Action Plan, using a standard format. Typically issues to be addressed in the Airport Environmental Management Plan are:

- [ ] noise
- [ ] water quality/drainage
- [ ] groundwater quality
- [ ] air quality
- [ ] hazardous materials
- [ ] solid and liquid (including quarantine) waste
- [ ] site contamination
- [ ] ecology (avifaunal habitat and flora)
- [ ] aboriginal and heritage issues
- [ ] landscape (visual impacts)
- [ ] energy management
- [ ] traffic management.

Environmental Contingency Plans

The AEMM also requires the development of Environmental Contingency Plans, which constitute a significant aspect of due diligence.

IMPLEMENTATION OF THE FAC ENVIRONMENT MANAGEMENT SYSTEM

Development of the FAC EMS as described above was only the first step in changing the management of environment issues at Federal Airports. Implementation of the system is more challenging and is ongoing within the FAC.

Following its endorsement by the Board, Airport Managers were asked to implement the EMS at their airport by developing an airport-specific Environment Management Plan.

Several months later, Corporate Head Office reviewed progress with the implementation of the EMS, with a view to establishing whether airports needed assistance with the task. It was evident that there had been little progress with the development of the Airport Environment Management Plans, and that this could be attributed to the following.

- ☐ The environmental issues arising out of the recently completed environmental audits of the airports were specific (physical) and therefore more apparent. In some instances they were pressing, so Airport Management's attention was focused on rectifying these matters.
- ☐ The Airport Environment Managers were in the main the personnel who had been responsible for undertaking the recently completed environment audits. To them, rectification of the issues arising out of the environmental audits was a priority.
- ☐ Although the EMS had been developed in consultation and agreement with Airport Managers, the actions required to implement the system within the airport were not clearly understood.

In response to these findings, and to emphasise management's commitment to the EMS, a directive was issued that all airports were to develop an EMP within a specified time.

Shortly afterwards the opportunity of a biannual managers' meeting, attended by all Head Office and Airport Managers, was taken to raise awareness of the role and benefit to the Corporation of the EMS. Airport Environment Managers were similarly appraised at a Technical Conference, at which time the type of actions required to implement the EMS within each airport were reviewed.

By the end of 1994, all airports had developed an Environment Management Plan incorporating the elements specified in the AEMM.

Understandably the approach of the Environment Management Plans varies between airports, as does the detail. This may largely be attributed to differences in:

☐ management approaches at the airports, and
☐ resources and the allocation of responsibilities at each airport.

The most significant example of this is the very different roles which the Airport Environment Manager has at each airport. For example, general aviation and regional airports, such as Bankstown, Essendon, Alice Springs and Coolangatta, have fewer staff than the major airports and responsibility for environmental management at these airports tends to be centralised. The Airport Environment Managers are responsible, not only for the development of the Environment Management Plan, but also for actioning most aspects of the Plan. Amongst other things, they are responsible for ensuring that the environmental responsibilities of all staff are defined within their duty statements, assessing and providing the environmental training needs for all staff, ensuring that environmental management requirements are specified in all agreements with contractors, monitoring and managing the tenant activities, preparing and implementing all environment action plans, responding to complaints and reporting environment management activities to management.

In contrast, at primary airports such as Sydney, Melbourne and Brisbane, responsibility for management of the airport's diverse business is spread and, as a consequence, environment management responsibilities are delegated more widely. Here the Human Resources Manager is responsible for ensuring that the environmental responsibilities of all staff are defined within their duty statements. The line manager and the Human Resources Manager together assess and provide the environmental training requirements of staff and the Technical Services Manager ensures that contractors' environmental requirements are specified in their contracts. The Commercial or Property Manager is responsible for ensuring that environment management requirements are specified in tenants' leases and, together with the Technical Services and the Operations Managers, monitors tenants' activities in this respect. The Technical Services Manager reports environment management activities to Head Office Management.

In this case, the Airport Environment Manager is responsible for coordination and oversight of the Plan, but it is actioned by others.

Other noticeable variations in the Environmental Management Plans are the different roles of the proposed Airport Environment Committee, particularly at the larger airports. In some instances the Committee consists of Corporation staff only, and is the forum used to coordinate environment management responsibilities within the Corporation at the airport.

In other cases, the Committee has tenant representatives as members, as well as Corporation staff, and is the forum for coordinating environment management responsibilities throughout all aspects of the airport's business.

Clearly both approaches have merit. The challenge for the Corporation now is to use the variations in the environment management approaches of the airports to its benefit.

PROBLEMS AND BENEFITS OF IMPLEMENTATION OF THE EMS WITHIN THE FAC

The benefits of implementation of an EMS within the FAC by far outweigh the problems which have been associated with its implementation. For example, the most frustrating problem to date has been the initially slow response in the development of the Airport Environment Management Plans. Besides setting a target for implementation of the system, the solution was to improve understanding of the actions required to implement the system at the airports.

On the other hand, the benefits of implementation of an EMS within the FAC have been numerous.

☐ Most importantly the FAC, the Board and Managers at Head Office and the airports have, for the first time, an appropriate, corporate-wide system for managing environment matters and thereby demonstrating due diligence.
☐ Consultation with Airport Management during the development of the system has ensured ownership of, and commitment to implementation of, the system throughout the Corporation.
☐ Development of an Environment Management Manual to provide a performance specification for the development and implementation of the EMS within each airport has ensured a consistent standard for the management of environmental issues throughout the Corporation, while allowing individual airports the freedom to tailor the system to their needs. This in turn has improved ownership, and commitment to the EMS.
☐ The development of Airport Environment Management Plans has established long-term environment management objectives for all

environmental effects on the airports, which has assisted budgeting for these issues.

☐ Variation in the way in which environmental issues are managed at airports allows for the development of new and innovative management approaches which can be developed and adopted to the benefit of all. At the same time, where a management approach does not work well, other models are available for airports to adapt to their requirement.

☐ Responsibility for the management of environmental issues is generally more widely spread throughout the airports, and the Corporation, and as a consequence environmental management is now more integrated with airport operations. The perception that this responsibility rests with the 'Environment Officer' or the 'Technical Section' is slowly disappearing.

☐ Development of airport-specific Environment Management Plans means that each airport has a fully functional EMS which, with only minor modifications, can continue to operate independently when the airport is privatised.

The last point has become particularly relevant as in May 1994 the Australian Government announced that it intended to privatise all Federal Airports.

FUTURE CHALLENGES

The aim of the FAC with the introduction of the EMS, goes beyond simply ensuring that environmental issues are consistently and systematically managed, by integrating them into the everyday business of the airport. It is also aimed at bringing about a cultural change within the organisation, to the extent that all individuals, employees, tenants and contractors, become aware of, and take personal responsibility for, environmental management.

This is the major environmental management challenge for the Corporation of airports over the next years, and it will only be achieved through continuous communication and training at all levels. This challenge will be met through the continued commitment and support of the Board and senior management, which is vital to the success of any corporation's environmental aspirations.

Human health impacts

PROFESSOR JIM BRIDGES, DR LESLIE HAWKINS and
DR OLGA BRIDGES, European Institute of Health and Medical
Sciences, Sterling House, Surrey Research Park, Guildford, GU2 5RF,
UK

Evaluation of the overall effect on human health of building, or of
extending the activities of an airport is by no means simple.

A variety of factors associated with airports have been claimed to
impact on human health. Concerns have been raised regarding the
contributions from noise, air pollution and psychological stress.
However, positive attributes have also been identified, in particular the
beneficial effects of providing jobs for the previously unemployed.[1] In
the present paper we will be concerned primarily with the assessment
of the impact of airborne chemical pollutants.

It is important, in making an assessment of the effects of chemicals, to
distinguish at the outset between the perception of odour and true
health effects. There is a popular misconception that if an unpleasant
odour can be detected, then adverse health consequences are likely. In
reality there is little evidence that an unpleasant smell itself will result
in any sustained adverse health effects, nor is there a general correlation between the level at which the odour of a chemical can be
detected and the level that causes mild toxicity.

SOURCES AND TYPES OF POLLUTANTS

Airborne pollutants associated with airports arise from three main
sources: ground vehicle movement, aircraft movement and fixed point
emitters such as power generation plants, fuel stores etc. There is little
evidence that significant levels of any chemical which is unique to
airport activities, will occur. The main pollutants arising are combustion products such as nitrogen dioxide, particulate matter and
volatile organic compounds including xylenes, benzene and alkylhy-
drocarbons. The same pollutants are to be found in urban communities
around the world, whether or not there is an airport in the vicinity. The
levels of these chemicals may of course vary between communities
with and without an airport.

IDENTIFYING SAFE LEVELS

To make a judgement on safe levels requires a detailed knowledge of
the toxic properties of the chemical under consideration and how
these properties vary in nature and intensity as the exposure level is
changed. Toxicologists generally divide chemicals (including airborne
pollutants) into two categories: (i) those with a threshold level below
which toxicity will not occur and (ii) those with no obvious threshold
level. The great majority of airborne pollutants appear to be in category
(i) but some carcinogens may be in category (ii). For pollutants such as
nitrogen dioxide and ozone, which have threshold levels for their
toxicity, it is possible to set standards or guide (guideline) values.
These are established using data (both human and animal) which
identify the threshold level, and building in a safety factor to allow for
differences in individual sensitivities of the human population. For a
number of pollutants two forms of standard are set, namely a short
term (acute) and a long term (chronic) value. The former is used as a
regulatory tool against sudden, but short lived, excursions from background levels. The latter is used to identify acceptable levels over a
period such as a year. To date the UK has only adopted four standards;
for nitrogen dioxide, sulphur dioxide, ozone and lead.[2] In comparison,
WHO has proposed guidelines for some 20 chemicals which it
recommends to aid governments in their public health strategies.[3]

The UK government is currently considering proposals for standards
for benzene, 1, 3-butadiene and PM_{10} (particulate matter of less than
10 μm). For other air pollutants the most widely used approach is to
take the workplace standard and divide it by an additional safety factor
(which may range from 10–1000) to provide an interim air quality
standard (iaqs).[4]

ASSESSING THE HEALTH EFFECTS
Existing activities

To assess the possible adverse effects from pollutants arising from
existing airport activities two broad approaches may be adopted.

☐ Comparison of airborne levels of particulate pollutants around the
airport against existing standards or iaqs values. The first requirement is to decide for which pollutants monitoring data are
required, and what their reliability and relevance to the airports

activities are. In the UK, routine monitoring data of this kind is often very limited indeed and the proposal to extend an airport's activities is often the initial trigger for a range of monitoring data to be gathered. Where such data have been generated, the contributions from the airport, although significant, are rarely very substantial.

☐ Examination of the health of the local population and/or workforce against that of a similar group not exposed to emissions from an airport.

The information available may be confined to health records over a period of time or may also involve functional measurements (such as respiratory function tests). Although, in principle, such studies are the ideal way of assessing whether or not an airport's activities are responsible for ill health, in practice such studies are fraught with problems. These include: imprecision in identifying health trends, the danger of inappropriate identification of a matching population, the somewhat diffuse nature of air pollution associated with airports and the consequent difficulty of linking any health changes to actual pollution. The airport workforce health records tend to be more reliable, though it is normally difficult to evaluate the relationships between health and exposure levels of particular pollutants. There is no reliable evidence that the profile of ill health of airport workers for respiratory diseases, for example, is higher than that of workers elsewhere.

Studies of the specific activities which might have potential health impacts have not been applied widely to airports although where they have, e.g. Schipol Airport, no obvious concern has been identified.[5]

Planned new developments

One, or a combination of, the following strategies may be adopted.

The most widely used strategy is based on modelling of anticipated levels of individual pollutants. The modelling should be directed to the identification of a realistic worst case scenario, with the levels of each pollutant of concern under both acute and long term emission conditions. The approach is to assume that a hypothetical individual lives permanently outside the airport at the point of the maximum level of each pollutant. It is also assumed that this individual is sensitive to each pollutant. The rationale is that if calculations reveal that this hypothetical individual will not suffer ill health then members of the local population, who will be subjected to a less extreme exposure, will also be fully protected.[5]

To achieve this a number of important decisions must be made. For the modeller, there is the challenge of which assumptions to employ, taking into account issues such as nitric oxide to nitrogen dioxide conversion, changing quality of road vehicle and aircraft fuels, and emissions with time. The toxicologist must identify the appropriateness of the iaqs values to use for those chemicals for which no national or international ambient air value exists. There is a danger that if too much conservatism is adopted in each decision, the final assessment will be totally unrealistic. In principle modelled data should be verified against actual data but in reality this is often not practicable.

Modelling ground levels of pollutants could include not only the impacts of airport expansion elsewhere but also major road developments associated with the expansion. In practice, to date, the quality of data available on which to base such judgements is generally very poor. This situation needs to be remedied.

There is much public concern regarding the impact of vehicle emissions on public health,[6] particularly asthma. It remains to be demonstrated whether vehicle emissions can cause asthma, although it is probable that exposure to high levels of vehicle pollutants can exacerbate a pre-existing condition. Previously, the vehicle pollutants of most concern were considered to be nitrogen dioxide and (indirectly) ozone. PM_{10} particles from diesel vehicles are now gaining similar prominence.

There is little evidence in any of the modelling studies associated with the further developments of UK airports that the proposed increased activities will lead to a significant enhancement of the levels of any of the pollutants assessed. Nonetheless, such evaluations of pollution trends associated with airports have stimulated an improvement in the quality of data on existing levels of atmospheric pollutants and the effects these may have on the health of various communities.

REFERENCES

1. Black Report.
2. *National Society for Clean Air and Environmental Pollution Handbook*, Ed Murley, Brighton, 1994.

3. WHO Air Quality Guidelines for Europe, 1987.
4. Van de Anker, I.M., Van Velze, K. and Onderdelinder, D. *Air pollution and Amsterdam Schipol Airport.* National Institute of Public Health and Environmental Protection, Billhoven, Netherlands, 1991.
5. Bridges J.W., Evidence presented to the Public Inquiry on the proposed second runway Manchester Airport, June 1995.
6. Royal Commission on Environmental Pollution. Eighteenth Report: *Transport and the Environment*, HMSO, London, 1994.

Conclusion

DR CALLUM THOMAS, Head of Environment, Manchester Airport Plc

Aviation is a comparatively young and highly regulated industry. It has grown very rapidly in the past two decades and therefore has evolved with an awareness of the environmental implications of its growth and operation. This has led to the development of new technologies and new, more environmentally acceptable operational practices. The industry does, however, retain certain legacies from the early years of aviation, in particular the impact of groundwater contamination.

As societies develop, increasing affluence is associated with increasing demands for air travel and at the same time, a lower tolerance of disturbance and environmental pollution. In addition, the continuing rapid growth of aviation, particularly in the developing world, is beginning to raise concerns about the global implications of the industry. This is because in many areas, growth is so rapid that it is offsetting the benefits of the technological improvements made in the recent past.

Aviation, however brings considerable social and economic benefits to society, both locally and globally. These benefits manifest themselves in a variety of ways. Aviation encourages global trading and investment, travel for leisure and education and cultural development.

The challenge for those involved in the industry is to ensure that the environmental impact of aviation is kept to a minimum so as to enable the industry to bring maximum benefit to society.

In recent years we have seen airports evolve their environmental control processes into comprehensive environmental management systems which influence every aspect of their operation.

Today, serious consideration is being given to understanding the concept of sustainability. It is now clear to all involved that the way in which environmental issues are handled holds the key to the future success of not only individual airports and airlines but indeed the aviation industry as a whole.

This conference has attempted to bring together a number of experts in the field of environmental management and sustainable mobility to describe new developments, to discuss new approaches and to share current best practice with others working in the field. This type of open discussion and debate is critical if the industry is to evolve and become more sustainable. In the final analysis, the achievement of sustainability is in everyone's own self interest.